CW01496176

5S

A Visual Control System for the Workplace

Edward Moulding

authorHOUSE®

AuthorHouse™ UK Ltd.
500 Avebury Boulevard
Central Milton Keynes, MK9 2BE
www.authorhouse.co.uk
Phone: 08001974150

First published by AuthorHouse 3/29/2010

ISBN: 978-1-4490-2977-7 (sc)

This book is printed on acid-free paper.

Contents

Introduction

The management of quality is a major issue for both private and public organizations. It requires the efforts of everyone – from the Chief Executive Officer (CEO) to the lowest-paid member of the organization. As a valued member of your organization, you can contribute your knowledge, support and participation, all of which are essential elements in ensuring success at implementing any improvement measures.

By the time you have finished reading this book, you will have a sense of how 5S activities will give you tangible benefits [1] –making your workplace a safer, cleaner, and more enjoyable place to work.

The aim of this book is to give a clear framework that can actually be used to improve the delivery of quality by giving practical guidelines on the first steps of process management.

---

1 Service Quality surveys imply an important message. Respondents' choices are summarized as follows: Appearing neat and organized, being reliable – do what you say you are going to do with assurance – and having knowledge and the ability to convey trust.
Valarie A. Zeithaml, A. Parasuraman, Leonard L. Berry. Delivering Quality Service, (New York The Free Press 1990), 27

Appearance Matters

Appearance matters when an organization or a brand is being judged. After all, appearance helps to claim a place in the minds of the organizations' stakeholders; it's about reputation. It implies a system of assumptions based on underlying values (culture).

Barbara Cartland once said: "I hate the type of boorish individual who, I am told, has hidden beneath such a rough surface a heart of gold. Quite frankly, I am a busy person and I have no time to dig."

This statement rings true of organizations of all types: for-profit, public-sector and not-for-profit organizations.

The customer or prospective customer has no time to dig; the organization must demonstrate it is good from the very first moment. Organizations are like living organisms; those that are fit and healthy move and change with environmental conditions through innovation and by developing a flexible relationship with their environment.

In an environment of competitiveness, companies – in manufacturing or service industries – must explore methods to operate more efficiently to meet the demands of their external and internal customers , so they often turn to programs and business-improvement tools such as Six Sigma, Total Quality Management (TQM), ISO 9001, ISO 14001 Environmental Management, Hazard Analysis and Critical Control Points (HACCP), Balanced Scorecard, Lean Operations and Customer Relations Management (CRM) to stay competitive, with an aim to achieve customer loyalty.

Achieving success in any of these programs requires a focus point – a philosophy that will carry the organization forward into improved productivity and efficiency.

One such tool in which staff at every level can participate to enhance the organization's ability to improve is 5S, which is an easy technique that uses the same language for the whole establishment. Everyone from top management to the front-line people can participate in the 5S program. By its nature, 5S will lead you to programs such as Six Sigma, TQM, ISO 9001, ISO 14001 Environmental Management and other quality standards, or Balanced Scorecard, lean and seamless systems, CRM, and Corporate Social Responsibility (CSR).

While 5S may seem to have an outside-in perspective t (that is, it allows organizations to see themselves through the eyes of a customer, supplier or others not in its employ with greater objectivity), this often means abandoning yesterday's perspective and embracing a new reality. This is only partly true, as 5S also relies on the development of competencies mainly by evolutionary learning – an inside-out perspective – that is based on the idea of *kaizen*, or continual improvement. This philosophy implies that small, incremental changes routinely applied and sustained over a long period result in significant improvements. The *kaizen* strategy aims to involve workers from multiple functions and levels in the organization in working together to address a problem or improve a process. The team uses analytical techniques, such as value stream mapping and the five whys to identify opportunities and eliminate waste in a targeted process or work area.

Chapter 1
Concepts and Overview

- ➢ **What Is 5S?**
- ➢ **The Reasons for 5S**
- ➢ **5S Is Proactive**
- ➢ **Definition of the Terms Used in 5S**
- ➢ **Who Should Be Involved?**
- ➢ **Resistance to Change**
- ➢ **Benefits of 5S**
- ➢ **Summary and Overview of the Program**
- ➢ **Points to Ponder**

What Is 5S?

The 5S program focuses on organizational cleanliness and standardization to improve profitability, efficiency and safety by reducing waste of all types. It gives organizations the five keys to a total-quality environment.

The 5S philosophy was born in Japan within the *5 Pillars of the Visual Workplace* by Hiroyuki Hirano and *The 5Ss: The Five Keys to a Total Quality Environment* by Taashi Osada

The name "5S" comes from five Japanese words all beginning with S. They are:

S1	Seiri	Sort	(organization)
S2	Seiton	Set in order	(neatness)
S3	Seisou	Shine	(cleaning)
S4	Seiketsu	Standardize	
S5	Shitsuke	Sustain	(discipline, training)

At its most basic level, it can be seen as housekeeping.

The Reasons for 5S: Why Would We Need a Special Program for Housekeeping, and What is it?

We often only worry about what we see, but what lies beneath the surface could be ten times more valuable. What's worse is that we may not think the things we cannot see pose any problems. The invisible elements of the business, such as the overstock of material, are like an iceberg: there is far more that we cannot see below the surface than what we can see above the surface. Even the visible becomes invisible with time because we step over the waste each day.

We often tend to ignore problems because they've been there since the start of the business, the dawn of time, or we regard them as part of our operating system and dismiss them by saying, "It's the way we do things."

Examples of the results of ignoring problems are ineffective meetings, lack of discipline, excessive stock, hollow projects with no objectives, no performance measurement, no clear understanding of processes and so on. In short, processes that have outlived the organization's original purpose by no longer fulfil a need. These practices have been with us so long that we not even recognize them as threats or constraints to our operation.

But why would a special program for cleaning help relieve these problems? It is because 5S is more than just cleaning, or housekeeping, from time to time. Housekeeping implies cleaning after a mess has been created – a reactive approach.

Those who apply 5S principles just as a housekeeping program, one only for aesthetics, will miss the true value of 5S to improve the operation's performance.

The idea of "7 Wastes" described by Taiichi Ohno – transport, inventory, motion, waiting (idle time), over-production (overstocking, high levels of obsolescence), overprocessing and defects (add to these waste in human resource, energy and water resources, materials, customers time and customer defections) – is a useful way of looking at problems, as it allows organizations to catalogue and categorize them, enabling them to focus attention on only those areas that require improvement. However, Ohno's 7 Wastes is not a tool to rectify the cause of the waste in the first place – the 5S method is that tool.

5S Is Proactive

The 5S approach is proactive: orderliness is designed into the processes by the designation of storage locations for tools, supplies, records and materials. First, these locations are selected and designed based on a well-thought-out, rational approach. Then preventive measures are implemented to ensure that the work area stays clean, thus reducing the need for conventional housekeeping.

A 5S project should be one of the early steps in a comprehensive range of initiatives, such as Lean Manufacturing (or, as this book considers other applications, Lean Operations is a more appropriate term), as we strive to reduce waste of any type and manage the necessary changes in business caused by changes in the marketplace or organization and enhance the customer experience both internally and externally.

Definition of the Terms Used in 5S

Before proceeding, let's take a short review of these terms, together with their definitions and objectives.

Definition	Objective
S1 – Sort (Organization)	To decrease waste and loss
S2 – Set in Order (Neatness)	To increase efficiency
S3 – Shine (Cleaning)	To observe, inspect and correct
S4 – Standardization	To reduce variables
S5 – Sustain (Discipline)	To maintain and train

Who Should Be Involved?

Adopting 5S at every level of staff will strengthen the daily management of an organization, as it is an easy technique to apply and it speaks the same language across the whole organization. Everyone – from top management to front-line staff – needs to participate to make the program a success.

By looking at the line of command in an organization chart, the 5S program can be divided into three levels,

while in terms of job content it can be divided into three
levels as it applies to job content.

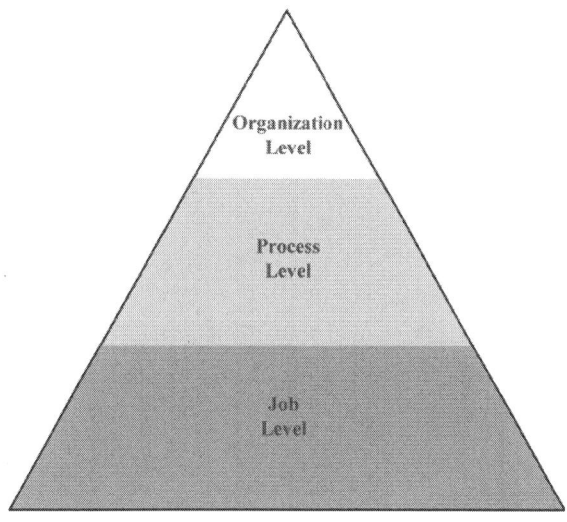

Figure 1.01

We will be exploring these concepts in the coming
chapters.

Resistance to Change

Any organization introducing a change strategy
such as 5S is likely to encounter some amount of resistance that will need to be addressed through training.
Demonstrating the benefits of the program and the ways
to overcome resistances.

Resistance to change often appears as:

➤ Fortress management style

Managers and staff working only with their own departmental objectives and not collaborating with others, this can creating barriers for communication across the organizations

And comments like:

➤ What's so special about Sort and Set in Order
➤ Why clean, when it gets dirty again
➤ 5S will not boost output
➤ We are too busy
➤ We are doing all right; we've always done it this way

Resistance to change and the seven wastes are some of the main constraints (bottlenecks) in any system.

Benefits of 5S

Some of the benefits to the staff and the organization are:

➤ Gives staff an opportunity to provide creative input
➤ Makes the workplace more pleasant to work in
➤ Adds job satisfaction
➤ Reduces defects to improve quality
➤ Reduces waste to lower costs
➤ Reduces delays to improve reliability of deliveries

- ➢ Promote safety by reducing injuries
- ➢ Improves availability of equipment by reducing breakdowns
- ➢ Increases customer confidence, trust and loyalty and reduces customer complaints
- ➢ Builds a culture of continuous improvement and team spirit

Summary

The key to success is not only to implement consistent methods to improve efficiency, but also to make the 5S practices a vital part of the organization's culture.

Improvements based on industrial engineering disciplines can be applied to offices, retail outlets, repair workshops as well as manufacturing operations, and all these will benefit from the application if 5S practices.

One of the key benefits of 5S is that it sets the foundation for continuous improvement. The 5S program improves profit by instilling a culture of quality and safety in the workplace. It does this by reducing waste of all types – time, material, supplies and maintenance – and simplifies the working environment, thus increasing morale and, in turn, improving efficiency and productivity.

Points to Ponder

Take a few minutes to think and answer some of the following sample questions[2]:

> ➤ What are the issues (bottlenecks) facing our organization that may affect its future (SWOT analysis)?
> ➤ Think... does your organization practice any 5S activities already?
> ➤ Visualize your workplace; Think of one area that would benefit from cleaning
> ➤ How could you promote carrying out this activity?
> ➤ What information is needed to fully understand these concepts and how can you get the information?

---

2 Mind mapping can be a useful way of taking notes.

Chapter 2
S1 – Sort
(Seiri)

- ➢ What Does Sort Mean?
- ➢ Organizational Level Policy
- ➢ Process Level
- ➢ Job Level
- ➢ Cleaning the Workplace
- ➢ The 48-Hour Rule
- ➢ Implementation Steps to Identify Unneeded Items
- ➢ Red-Tag Strategy
- ➢ Accumulation of Waste
- ➢ Summary
- ➢ Points to Ponder

What Does Sort Mean?

Sort means to examine everything in each work area and storage area, including computer systems, and then evaluate the need for each item. Sort calls for the elimination of unnecessary items that have collected around the work area. Why is **Sort so** important? It is important because when this first step is implemented well, it sets the tone for the rest of the steps in the **5S** process by reducing the constraints on the flow of work, thereby improving the work environment to use resources such as space, time, energy, people and money more efficiently.

Organizational Level Policy

At the policy level, Sort involves systematically setting and distributing policies and checking the policies, process instructions, work instructions, flow charts and organization charts so that they are clear, not overlapping, and up to date and that they support the vision of the organization by focusing on making it lean.

Process

At the process level, Sort involves managing the workflow for services, production and customer care processes so that there is no overlap between processes and that they run efficiently, using only necessary steps, and focus on short, optimal, measurable procedures.

Job

At the job level, Sort involves doling out jobs according to their necessity, which will prevent waste and loss.

Cleaning the Workplace

Cleaning means to go through everything in each work area, keeping only what is necessary. Materials, tools, equipment and supplies that are not frequently used should be moved to a separate storage area, and items that are not used should be discarded. When discarding items, workers often have a strong emotional attachment to things and processes, so handle this process with sensitivity. This can be done using tools such the Delphi Technique.

The 48-Hour Rule

A lot of what is in the workplace, workshop, factory, office, retail shop or showroom can often fall into the "might be used someday" class.

It's often comfortable to have these items around. This is when to apply the 48-Hour Rule: Figure 2.01

If it's not going to be used in the next 48 hours, then it does not belong in the work area.

48 Hour Rule

If it is not going to be used in the next 48 hours, then it does not belong there.

Figure 2.01

The steps to identifying what is not needed:

- ➢ Define what is *needed*
- ➢ Define what is *not needed*
- ➢ Dispose of the items
- ➢ Take action

Implementation Steps to Identify Unneeded Items

It is not always so simple to identify exactly which items are required. While the staff will seldom give very much consideration to what is needed or how to separate out items, the management often sees waste without recognizing it. To recognize waste and understand the ways to deal with it, you will need to develop the capacity to assess your workplace. It is important to realize that this requires a structured methodology to ensure the involvement and commitment of all personnel so that the benefits of the 5S program can accrue. This will encourage a systematic process for auditing the workplace in future for other quality programs – ISO 9000, ISO 14001. Such an approach can be obtained by applying the 'Red-Tag' strategy.

Red-Tag Strategy

This simple tool gives us the method for evaluating items' utility so that we can deal with them in an appropriate way by asking just three questions:

- ➢ Is this item needed?
- ➢ If so, is it needed in the quantity on hand?
- ➢ If yes, do we require it to be located here?

Defining What Is Needed

When considering what is needed in a workplace, look beyond the core tools and equipment that are used and also look at your materials – supplies, paperwork, processes and even computer files and programs. Take this as a great opportunity to consider if you have the correct quantities of each item. Has your workplace become a PC zoo?

Defining What Is Not Needed

If it is not needed for a task that supports core processes —making a product or providing a service — such as safety items, it most likely should be kept outside the direct working area.

Dispose of Items

Apply a red tag using a standard set of guidelines to those items to be removed . (We will return to setting the criteria for this a little later.) Then record data to determine how frequently each item is used, and store the least frequently used items the farthest from the work area.

Take Action

Once the disposition task is complete, it is time to take action to clear the red-tagged items to a holding area. A holding area is an area set aside to store red-tagged items that need additional evaluation. Here the red-tagged items can be examined in detail to confirm that they are not to be kept in the work area. This can be done by an audit team who can use tools like the Delphi Technique to overcome those strong emotional attachments.

Processes for Red-Tagging

The process for the red-tagging project can be broken down into seven discrete steps:

- ➢ Launch the project
- ➢ Identify the targets
- ➢ Set the criteria
- ➢ Design and make the Red Tags
- ➢ Attach red tags
- ➢ Evaluate
- ➢ Implement Documentation

Launch the Project

For this step, the organizational level of the company has to coordinate and start the project. This involves organizing the project team – bringing together a group of people from different parts of the organization. This can be beneficial since they will often see things from a fresh prospective. It is also an opportunity to develop and share skills. The team's duties are to:

- ➢ Arrange any supplies that are required,
- ➢ Set the schedule to perform the task,
- ➢ Plan and set aside the red-tag holding area,
- ➢ Set the target for items to be red-tagged,
- ➢ Plan the disposal of red-tagged items.

Identify the Red-Tag Targets

This requires identifying two things:

- ➢ The specific types of items for evaluation,
- ➢ The physical areas where the red-tag action will take place.

When considering the types of items, include document files, catalogues, equipment of all types and inventory. Inventory should be separated into items that are in progress and warehouse inventory.

When identifying and defining the area for evaluation, it is recommended to focus on a small area and carry out each task to completion before moving on to the next rather than trying to complete a larger area and then fail to fully complete the task.

Red-Tag Targets

Physical Areas

- Floors
- Walkways
- Stairs and stairwells
- Fire escapes
- Desks (tops and insides)
- Walls and notice boards
- Shelves
- Warehouses
- Tool boxes
- Cabinets
- Customer areas
- Car parks
- Loading areas

Inventory

- ➢ Stock (e.g., excess inventory)
- ➢ Semi-finished products
- ➢ Work in progress
- ➢ Supplies in the work area
- ➢ Scrap and rework
- ➢ First-time yield
- ➢ Office stationary
- ➢ Equipment (e.g., damaged electrical wiring)
- ➢ Computers (e.g., PC zoos, printers)
- ➢ Machinery (e.g., jigs, dies)
- ➢ Lighting, water, heating, air conditioning equipment
- ➢ Old material (e.g., cleaning rags)
- ➢ Old, out-of-date documents and posters
- ➢ Fire hazards

Exercise

Ask yourself what could be targeted in your own organization or workplace and then identify the following:

- ➢ Types of items
- ➢ Physical areas

Set the Criteria

We have already spoken of the most difficult task surrounding the red-tag process – defining what is needed and not needed. This problem can be managed by establishing clear-cut criteria for what is needed. It is recommended that 5S teams for each organizational or operational unit set its own standards. There are just three factors to consider for doing so:

➢ Usefulness: If the item isn't needed, dispose of it.
➢ Frequency of use: If the frequency is low, store it away from the work area.
➢ Quantity needed to perform the work: This is a good time to be thinking of Just-in-Time (JIT) techniques.

Design and Make the Red-Tags

Take time to design the various types of information that you want to record. Figure 2.02 They should support the organizations' documentation and reporting processes and aid you in the measurement and mapping of the project's progress. They may include:

➢ Category
➢ Item name and number
➢ Quantity
➢ Reason why it has been red tagged
➢ Department responsible for the item
➢ Value
➢ Date

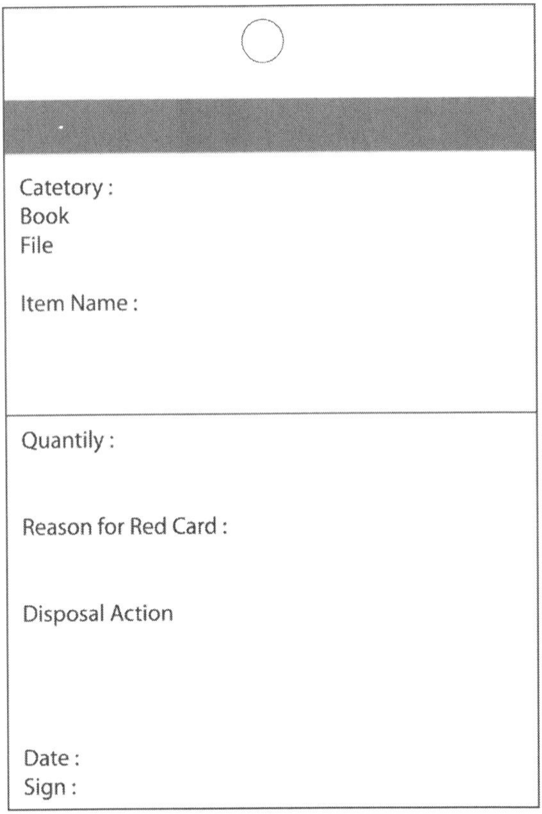

Figure 2.02

Attach Red-Tags

This task should be done as quickly as possible – a short, powerful event performed in one or two days. Some practitioners use a technique called Big Cleaning Day to clear out as much as possible as quickly as possible, in a big blitz. However, it comes with a word of caution: if you use this technique, make sure that the project does not stop there with people already feeling satisfied.

At this stage tag all items you question without over-evaluation of what to do with them. Further evaluation will come in the next step.

Evaluate

Now that items in the target area have been red-tagged, the next step is to evaluate the items in detail. Using the criteria established earlier, we can determine our options for action, which are:

- ➤ Keep the item where it is,
- ➤ Move it to a new location,
- ➤ Hold it in a special holding area for further evaluation,
- ➤ Dispose of the item.

Implement Documentation

Documenting the actions taken will allow you to track information as the process takes place and to measure the improvements and savings made. Digital photographs are a useful resource, particularly for training in 5S techniques. Also a simple database computer spreadsheet can be created.

Accumulation of Waste

Certain types of waste tend to accumulate in predictable places, some of which are listed the Red-Tag Targets

Once evaluation is complete, give some thought to disposal methods.

Disposal Methods

Scrap/Throw away	Sell for scrap; make sure that the disposal method complies with environmental regulations
Sell	Sell off surplus to other organizations
Return	Arrange a buy-back of excess stock (stock adjustment); this will improve your ROI
Distribute	Distribute the items to other parts of the organization where they can be used
Give away	Donate the item to a charity, school or college that can make use of it
Obsolesce	Ensure there is a policy and process for dealing with obsolete stock

Summary

We discussed the meaning of Seiri – Sort and the responsibilities at the various levels of an organization's structure. Then we looked into how we can implement Sort – by removing all the unneeded items from the workplace. With the "when in doubt, throw it out" philosophy, problems and annoyances in the workflow will be reduced. You will then have taken the first step towards a lean organization.

Use the red-tag strategy for evaluating what the organization needs and does not need. Think if it's not going to be used in the next 48 hours, then it does not belong in the work area

This simple method's seven steps for identifying potential waste is your starting point. It will produce significant results.

The best way to carry out red-tagging is to target a whole area, then apply red tags to items in one or two days. Also, set a target for the number of red tags to be used.

Be alert that certain types of waste tend to accumulate in predictable places.

Consider your options for disposal methods and document your progress.

Take before-and-after photographs. These will be useful in the next step of 5S.

Points to Ponder

Before moving on to the next step, Set in Order, take some time to reflect on the results of your efforts.

- ➤ What problems are occurring in your workplace because of an accumulation of unneeded material?
- ➤ List five targets (material and areas) for red-tagging. What criteria would you use?
- ➤ What did you learn?
- ➤ What outcome was particularly useful?
- ➤ Did you have all the information you needed to complete the task? And if you did not, how could you get it?

If Sort is done well, it produces impressive results. The work environment, quite literally, changes overnight. Don't forget to note down your findings.

Chapter 3
S2 – Set in Order
(Seiton)

- ➢ What does Set in Order Mean?
- ➢ Organizational Level
- ➢ Policy
- ➢ Process
- ➢ Job
- ➢ Importance of Set in Order
- ➢ Visual Control
- ➢ Ergonomics – Principles of Motion Economy to Reduce Waste
- ➢ The 5S Map
- ➢ Signboard, Painting, Colour Coding and Outlining Strategies
- ➢ Summary
- ➢ Points to Ponder

What Does Set in Order Mean?

Now that your workplace has been Sorted (cleared of waste), it is time to move on with the next step in the process – Set in Order. **The key objective here is to put everything in its place, properly identified and labelled.**

In a nut shell, Set in Order means neatness.

Set in Order can only successfully be implemented when the Sort step is complete. There will be little benefit from the 5S project if many unnecessary items that are still in place. Sort and Set in Order work best when implemented together, as Set in Order is much less effective without the Sort process.

Organizational Level
Policy

At the policy level, Set in Order encourages neatness and speed by supporting necessary resources, operating systems and technology in order to respond quickly to customers' wishes or to rapidly adjust to changes in the business environment.

Process

At the process level, this step creates and improves working processes and procedures so that each step links smoothly to the next. This applies equally to service and manufacturing processes. The objective is to emphasize meeting customer demands promptly or adjusting to changes quickly.

Job

At the job level, the task is to increase performance and efficiency by arranging tools, equipment and document files in a proper order that enables quick retrieval.

Importance of Set in Order

Set in Order focuses on effective storage systems and systematic organization methods by arranging items in such a way that they are easy to find for anyone who uses them. **The keyword here is "anyone."**

Neatness Will Bring Speed

Dr. Shigeo Shingo, one of the engineers responsible for the Toyota Production System (TPS) and a recognized JIT authority, identified seven wastes as targets for continuous improvement. According to him, systematic organization with designated locations will eliminate many types of waste, such as:

- ➢ Motion
- ➢ Overproduction
- ➢ Waiting time (uneven workloads)
- ➢ Collection and returning time (transportation)
- ➢ Excess inventory
- ➢ Defective products
- ➢ The process itself (why it is needed)

Further types of waste are:

➢ Unsafe working conditions and practices
➢ Searching time
➢ Under unitization of man power
➢ Pollution
➢ Energy and water

All operations suffer their fare share of time and costs caused by these examples of waste. By attending to these wastes, improved efficiency – both environmental and social – can be achieved.

The Benefits of Standardization

Most people know the Battle of Waterloo [3] as one of those defining moments of European history, but few realize that British Standards played a decisive role in it.

The French cannons used cannon balls of different sizes. If your artillery unit ran out of cannon balls, then you were out of luck, as the other field artillery units could not replenish your supply.

In contrast, the British had specified that all cannons and cannon balls supplied to artillery units had to be of a standard dimension. As a result, all cannon balls could be used in any cannon.

---

3   Kit Sadgrove and Kogan Page, *ISO 9000 / BSI 7550 Made Easy*. (London 1994), 23

In today's language, this means higher productivity, less waste, less cost and less down time – all in all, a leaner operation. Keeping this lesson in mind, the British Standards Institution was formed a few years later.

Set in Order – The Core of Standardization

Before Standardization can be considered, order and neatness need to be achieved through attitude, awareness and culture to capture the essence of Set in Order. Figure 3.0.1.You cannot implement any form of Standardization without order, and the central tenet of this culture is process control.

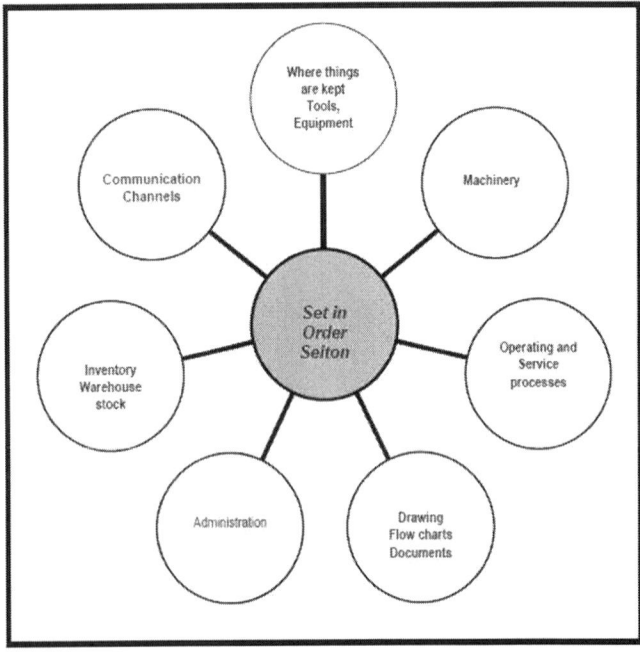

Figure 3.01

Visual Control

Visual control is any communication method used in the work environment that tells us at a glance how a task should be done. The information conveyed through visual control includes the identification of where certain items belong and how many items belong there (see Figure 3.02 an example of visual control to manage a tool room.

Figure 3.02

Systematic labelling allows for the easy return of any item to its proper place. The process should extend to your entire facility, including places where safety is an issue. In many cases you need to affix Right to Know (RTK) labels to containers or equipment (Figures 3.03 and 3.04) fire prevention and body parts rack & storing work Instructions.

Figure 3.03

Figure 3.04

Visual control management systems provide real-time information and feedback on the status of the operation. It is an organization-wide "nervous system" that allows all employees to understand how they affect overall performance.

Visual control is a total-quality-control technique that makes any problems clearly visible to all employees, allowing them to bring in resources to resolve the problem as it arises.

The objectives of visual control management systems are employee involvement and motivation, open communication, quality productivity improvements and faster decision-making processes.

Ergonomics – Principles of Motion Economy to Reduce Waste

Ergonomics is the study of work as performed by the human body. It can help us maximize output without causing physical harm to ourselves

The principles of ergonomics help us in deciding the best location for things to prevent injuries from repetition, overload or awkward body positions. This can be illustrated in the following two diagrams. Figure 3.05 and Figure 3.06

Zone 1 contains the most frequently used objects, and Zone 4 contains the least frequently used objects.

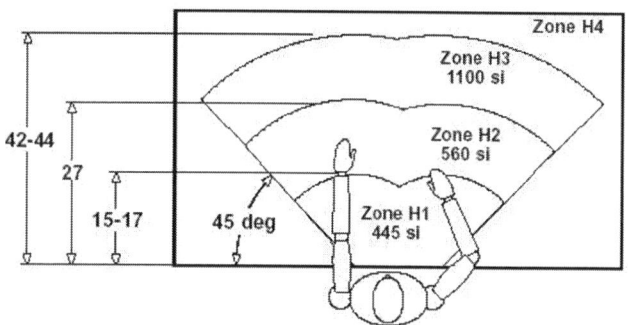

Figure 3.05

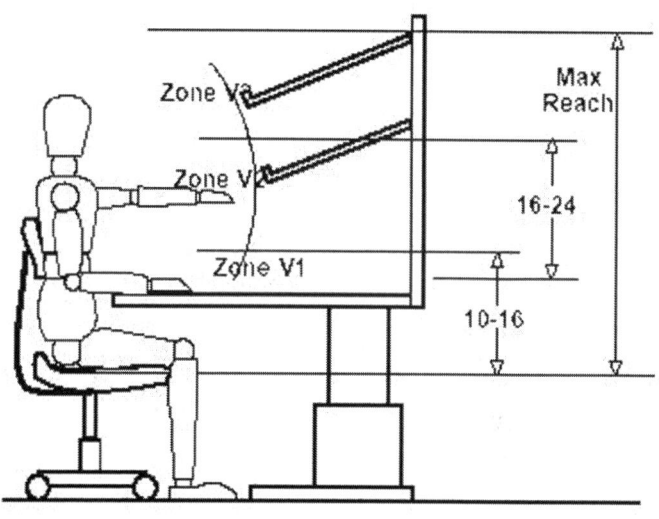

Figure 3.06

Adapted from: McCormick, 1964.

By locating items, equipment, machinery and tools in the best possible location, we can minimize waste. This is referred to as "**motion improvement**." Consider if it may be possible to find ways to remove waste by eliminating the whole operation; this is called "**radical improvement**. Figure 3.07

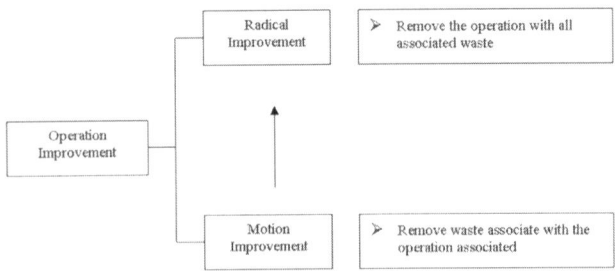

Figure 3.07

The 5S Map

The 5S map is a tool for evaluating orderliness and the current location of equipment. Using 5S mapping actually involves making two maps – a **before map** and an **after** map. While the before map shows the layout before you have implemented Set in Order, the after map shows the layout when you have implemented **Set in Order**.

Look carefully at the first map to see where congestion in the workflow arises and to determine how that waste can be removed. Experiment with layouts to see which one gives the best results. Continue to analyse the effects of the layout on efficiency and look for improvements in orderliness of the work layout.

Signboard, Painting, Colour Coding and Outlining Strategies

Once the best locations have been decided, these locations should be labelled so everyone knows where things belong and how many of each thing should be located there.

Signboards are often used to identify the following:

➢ Names of work areas
➢ Inventory location
➢ Equipment and tool locations
➢ Job instructions

Painting Strategy

Painting or using tape is a method to mark out areas. Common types of painting markers include:

➢ Carts and trolley storage areas
➢ Aisle direction
➢ Door range to show which way the door swings
➢ Tiger marks to show where not to store things, or to show hazardous areas

These marks can be colour coded with bright and standardized colours, as in the following examples:

➢ Operating areas: green
➢ Walkways: fluorescent orange
➢ Dividing lines: yellow

Colour Coding Strategy

Colour coding can be used to show the status of equipment that needs calibration by having operation periods or lubricates colour coded or for coding lubricates.

Colour coding can also be used for Separate bins that hold parts that have expiry dates or for dangerous goods.

Outlining Strategy

Drawing an outline around tools on boards provides an easy visual control, as any missing tools can be identified at a glance. Use this technique for areas that are for special purposes.

Summary

Here we discussed the meaning of Set in Order and its key objective: to have everything in its right place, properly identified and labelled. The responsibilities at the various levels of an organization lie with policies to encourage neatness so that items are easy to find, use and properly return.

Motion economy principles also help reduce waste. We also looked at some methods of visual control to achieve Set in Order: the 5S map, signboards, colour coding, painting and outlining the placement of specific items.

Visual control is a total-quality-control technique.

The main purpose of Set in Order is to achieve visual control as a communication method for systematically organizing and labelling items so workers can identify what belongs where and in what quantities

Set in Order is the core to Standardization, which will be examined in chapter 5.

Points to Ponder

- ➢ Think of examples of visual controls in your workplace, whether it is a workshop, office, school, hospital or retail shop.
- ➢ Think how visual controls could be used to enhance the work environment and improve performance.
- ➢ List examples of the seven wastes that you see in your workplace.
- ➢ What did you learn?
- ➢ What outcome was particularly useful for you?
- ➢ Did you have all the information needed to complete the task? If you did not, how can you get it?

Remember to record all your findings.

Chapter 4
S3 – Shine
(Seisou)

- ➢ **What Does Shine Mean?**
- ➢ **Organizational Level**
- ➢ **Policy**
- ➢ **Process**
- ➢ **Job**
- ➢ **Importance of Shine**
- ➢ **Inspection**
- ➢ **Resources**
- ➢ **Implementation**
- ➢ **Summary**
- ➢ **Point to Ponder**

What Does Shine Mean?

"Shine" means to clean, but more than ensuring spotless housekeeping, it is about preventing the environment from becoming dirty in the first instance. Shine also includes observing or checking for problems that may arise (inspection).

The first two parts of the 5S process we looked at are Sort and Set in Order. These provide a strong foundation for workplace improvement, driven not only by the skills and knowledge of workers but also the values of organization and simplicity.

There will be barely any long-term benefit of 5S if the materials and equipment workers use are not just dirty, but also in poor working condition, frequently breaking down.

Clean workplace conditions are also important to employees' heath, morale and safety. All of these factors impact a company's bottom line and return on investment. Maintaining a clean and well-kept work area is an important factor in promoting increased performance, as it demonstrates the competence of management and how seriously they take their work. Too much down time caused by hunting for tools, equipment breakdowns or waiting for work instructions will all sap the motivation of the workforce. Ultimately, down time also shows a lack of sound planning by management.

Organizational Level Policy

At the policy level, Shine means operating an organization so that it becomes a **visual factory**. A visual factory uses signs and pictograms at the workplace to communicate essential information on safety and about how equipment, materials, locations should look and give easily understood instructions on performing tasks

This enables management to quickly assess the organization's status, and to forecast potential problems and to find solutions to problems more easily.

Note: The term **visual factory** applies to any work area, office, showroom, retail shop, hospital, school and so on.

Process

At the process level, Shine involves regular cleaning and inspection. In the long term, Shine will save time because operational staffs can immediately identify problems and take corrective actions.

Job

At the job level, cleaning activities can be useful for monitoring the condition of equipment, giving workers the opportunity to rectify problems as they arise.

Importance of Shine

While Shine's obvious purpose is to make the work environment clean, it also allows organizations to get beyond seasonal cleaning and build a culture of daily cleaning and inspection so that work areas and their equipment are ready for use at any time.

Cleaning activities play a significant role in aiding efficiency and workplace safety, which is rewarded by improvements in employees' self-esteem and confidence, which, in turn, adds value to the overall organization.

Inspection

When we clean an area, it is natural to perform a certain amount of inspection. Therefore, once you have established the routine of daily cleaning, a systematic inspection process can be introduced. Thus stands the process

cleaning and inspection.

Cleaning and inspection will detect signs of minor malfunctions in equipment and processes. Management should not wait for machinery to make odd noises, computers to slow down or materials and equipment to be misplaced. Inspection can preempt problems and save unplanned down time.

Resources

Ensure that the right tools and cleaning materials or agents are available; there is little point in cleaning if it cannot be done successfully. Also pay close attention to health and safety regulations for use of equipment such as steam or high-pressure cleaning products and the use and storage of cleaning solvents. Storage takes us back to Set in Order, in which the key objective is to have everything in its place, properly identified and labelled.

Note: You may be required to provide special training in the use of equipment or cleaning agents.

Implementation

Daily Shine activities should be taught as a series of steps with rules to maintain discipline. This way a clean, safe and easier to manage workplace will be achieved.

Planning Steps

1. Determine targets

These targets can often be grouped into three types:

- ➢ Warehouses;
- ➢ Equipment, including tools, machines, workbenches and office furniture;
- ➢ Space, including common areas, floors, windows, lights and shelves.

2. Assign tasks

Everyone working in the unit or department is responsible for workplace cleanliness, even if the organization employs cleaners. There are three tools we can use when assigning Shine tasks:

> ➢ The 5S map, discussed in "Set in Order",
>   shows the Shine areas and their associated
>   responsibilities.
> ➢ Drawing up cleaning and inspection checklists to
>   show exactly what tasks are to be done
> ➢ A 5S schedule details who is responsible for Shine
>   when.

3. Clean and Inspect

Once the cleaning and inspection tasks have been assigned It is important to allow time these activities so they become a natural part of the daily work routine.

4. Prepare tools

> ➢ Apply Set in Order to cleaning equipment so it is
>   easy to find, use and return.
> ➢ Analyze whether the areas containing cleaning
>   equipment are properly marked.
> ➢ Examine whether the cleaning supplies are
>   replenished and have usage instructions marked
>   on them.
> ➢ Figure out whether there are sufficient and clearly
>   labelled waste bins

5. Shine

Be thorough when cleaning so that dirt is removed from corners and pillars and areas such as windows, doors and work surfaces. Use cleaning agents when sweeping and wiping do not sufficiently remove dirt.

Cleaning coincides with equipment inspection and evaluation. Ensure that proper care and maintenance take place. Look out for such things as oil leaks, unusual noises, vibrations and excessive heat. Determine a course of action, such as:

➢ Instant maintenance and repair, whenever possible
➢ Requested maintenance if the operator cannot determine the cause of the problem or does not have the skills to perform the maintenance,

Log the maintenance done. Also log any maintenance requested and completed of the work. Figure 4.01

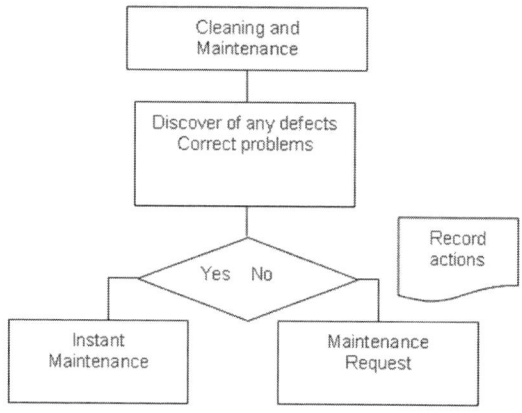

Figure 4.01

Summary

The benefits of a clean workplace become apparent in a short period of time. One of the key purposes of these activities is to keep equipment in good working condition so that it is always ready for use. Equipment that is kept clean performs more efficiently and requires less unscheduled down time as a result. When cleaning is implemented well, employees' morale will rise, safely standards will improve and productivity will increase.

Analyze the causes of dirt accumulation and look for any constraints that may prevent cleaning measures and find ways to overcome them.

There are five key steps to Shine: determine targets, assign tasks, clean and inspect, prepare tools and Shine.

Once these routines are established as periodical activities, they do not require much time and effort. Considering the short amount of time needed to maintain them, the results can be quite remarkable.

Points to Ponder

Before moving on to the next step, Standardize – Seiketsu, take some time to reflect on the results of your effort in the workplace. Don't forget to record your results with photographs, which can be used to note progress and set benchmarks.

- ➤ What problems were overcome at your workplace by applying Shine activities?
- ➤ What did you learn?
- ➤ Which outcome was particularly useful?
- ➤ Did you have all the information you needed to complete the task? If you did not, how can you obtain it?

Chapter 5
S4 – Standardize
(Seiketsu)

- ➢ What Does Standardize Mean?
- ➢ Organizational Level
- ➢ Policy
- ➢ Process Level
- ➢ Job Level
- ➢ Importance of Standardize
- ➢ Implementation
- ➢ Summary
- ➢ Points to Ponder

What Does Standardize Mean?

Standards are the specifications for performing work. Each standard describes the preferred method for performing the work, and it often describes the time the task should take to complete. At the same time, standards establish the boundaries of freedom within the task.

Organizational Level Policy

At the policy level, setting standards allows the organization's management to bring under control activities that would normally vary in results, or to keep activities relevant to the customers' needs. This stabilizes the daily management routine and will reduce waste and loss. As a consequence, management can devote more time to their main business policies: innovation and marketing.

Process

Set standards for processes offer supervisors the means to control work and, by doing so, reduce waste and loss in processes, thereby improving the organization's service or product.

Job

At the operational level, Standardize helps staff understand what is required of them and helps them get into the new habits that are part of the 5S program.

Importance of Standardize

The process of cleaning an organization's systems, established without any clear standards, loses effectiveness over time, and the good practices and progress made in Sort, Set in Order and Shine will soon be lost.

In chapter 3 we discussed the benefits of standardization as a result of Set in Order– using standards to help people work in new organizational routines. It's easy to slip back into old habits that we've done for years. As we feel comfortable with familiar work routines, breaking out of a culture of "we've always done it that way" attitudes that prevent us from getting better at what we do requires considerable effort, **not because standardization is difficult but because it is a new paradigm.**

Standards contribute to the overall quality and safety of our work and to the products or services we provide. They ensure compatibility, reduce unnecessary variety and increase the cost-effectiveness of processes and procedures.

Definition of Standardize

Standardize in terms of **5S** is defined as creating a consistent way of performing tasks. If viewed this way, Standardize is the result when **Sort, Set in Order and Shine** are well maintained.

The principles underlying the standards can be found in all engineering standards. They focus on using visual cues that are aesthetically appealing for the people doing work and their work environment.

Engineering standards are specifications for performing work. Each standard describes the preferred method for performing the work and also the time required to complete the task. Engineering standards can be applied to any environment, such as a retail shop, an office and a schools. They do not just apply to engineering workshops or factories.

Standardization in the context of 5S forms part of a general quality control procedure and results in the efficient performance of service delivery or manufacturing processes by improving environmental performance, such as improved energy efficiency, for instance. Thus, pollution and expenditure are reduced simply Standardize reduces waste.

These standards are based on the following five principles:

Five Principles

> An average worker who is capable and has had sufficient training;
> A consistent workplace that reflects average skill and effort;
> Prescribed work methods that are consistent, standardized steps performed in sequence;
> Specific working conditions – workplace layout, tools, materials and equipment;

➢ Capable supervision that gives adequate direction.

Implementation

Having established that Set in Order is the core of Standardize achieved by using visual controls as a means of communication for achieving systematic organization, the next step for implementation requires us to determine what the standards will be, who is going to do the tasks and how the tasks should be organized using the five principles for engineering standards.

The basic purpose of Standardize is to prevent setbacks – in Sort, Set in Order and Shine – and to make them a daily routine so they become simply **the way we work'.**

Ways to Implement Standardize

There are three steps to making Sort, Set in Order and Shine the **way we work:**

➢ Decide who is responsible for which activities
➢ Integrate 3S processes into the regular work pattern
➢ Audit how well 3S is being maintained

Responsibilities

Everyone must know exactly what they are responsible for doing and when, where and how it needs to be done. These are the basic elements – the building blocks

– of any process. The strength of a process depends on the weakest link in the chain within it. This means that the secret of successful 5S program is to ensure that each link, or step – Sort, Set in Order and Shine – is equally strong. Getting it right the first time at each step of the way is the byword.

Job assignments should be based on each person's own workplace, both for outsourced services and work performed at your premises, for Sort, Set in Order and Shine to have meaning.

Integration

Integrate 3S steps into the regular work pattern so that implementing them becomes part of the work culture. This will prevent conditions from reverting. This can be achieved by a thorough understanding of the elements of 5S processes so that everyone involved recognizes that they are part of the 5S program.

By understanding the principles of the process, we can recognize the importance of maintaining the strength of every link in the chain of the 5S process and reap the ultimate reward.

A powerful tool in our understanding of Standardize is a simple process diagram, such as the one shown in Figure 5.01

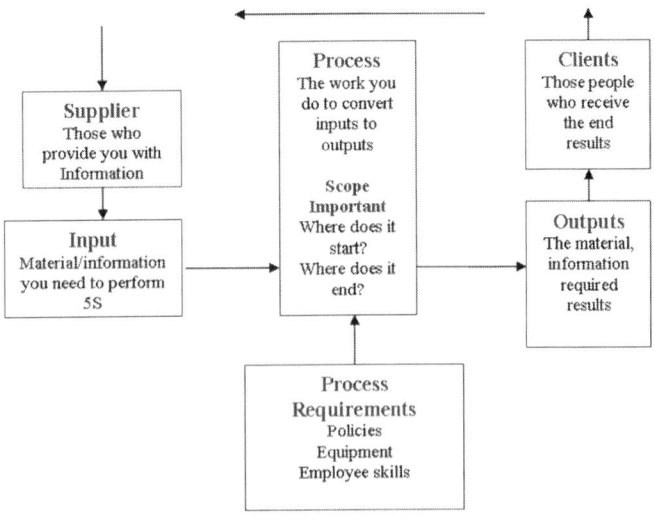

Figure 5.01

Adapted from *Understanding Total Quality Management in a Week'* **by John Macdonald, Institute of Management. (Hodder & Stoughton Ltd. London. 2nd Edition 1998),65**

Other tools and techniques available to us are **Visual 5S** and **Five-minute 5S'**. **Visual 5S** is a collection of control devices that tell us at a glance what needs to be done to maintain 5S processes. Visual 5S methods were discussed in chapter 3 as part of **Set in Order**. This approach makes the quality of 5S conditions obvious at a glance – anyone should be able to distinguish between abnormal and normal conditions just by looking at the work area. One can then make an objective decision from their observations about what is acceptable and un-acceptable.

The Visual 5S method triggers t**he Five-minute 5S** strategy, which is a set of corrective actions for 3S – Sort, Set in Order and Shine – once nonconformities becomes apparent. These nonconformities could be, for example overproduction, too much stock at the workstation, disorder, contamination and waste materials not cleared away.

The Concept of Prevention

If we find that tools, files and other equipment are not returned to their correct storage places, then we need to take care of this immediately. If there are spills on the floor or workbenches, we clean them up immediately. Making actions such as these part of the working culture is the foundation of Standardize. These actions will also help build a sense of responsibility in employees for themselves and their environment by avoiding negligence liable to harm them or others.

However, if the same problems keep arising, then it's time to start asking why the problem recurred in order to take the concept of Standardize to a new and higher level, this means solving the problem at its root cause.

The process of problem solving can be broken down into four steps:

- ➤ Understand the current situation
- ➤ Identify the root cause of the problem
- ➤ Develop an effective action plan

> Carry out the action plan, making adjustments as necessary to the plan until the problem is solved.

These steps form a single process. Before solving anything, first you need to realize there's a problem. Once you do, knowing the root cause isn't enough, think through how the problem can be fixed, then take the required actions to fix it. Problem solving is a combination of thinking and acting. Only doing one or the other will not get you anywhere.

Prevention
5W1H (Kipling Process)

This concept originated from the poem by Rudyard Kipling, "The Elephant's Child":

I keep six honest serving men
They taught me all I knew
Their names are What and Where and When
And Why and How and Who

By analyzing every part of our process and by defining the essence of the problems, we can work out the causes of the problems and find appropriate ways to solve them.

To achieve this, we must start by asking why. Until the underlying cause(s) is(are) identified, we might have to ask why several times before we get to the root of the problem.

Questions

What, Why, Where, When, Who & How.	
Place	Where is it done?
	Why is it done there?
	Where else might it be done?
	Where should it be done?
Sequence	Why is it done then?
	When might it be done?
	When should it be done?
Person	Who does it?
	Why does that person do it?
	Who else might do it?
	Who should do it?
Means	How is it done?
	Why is it done that way?
	How else might it be done?
	How should it be done?

The cause-and-effect diagram (C&E diagram) is another useful tool to further investigate the cause of a problem and to distinguish between the cause of a problem and its effect. The fishbone diagram figure 5.02 is one type of C&E diagram that can be used to brainstorm each cause and its effect

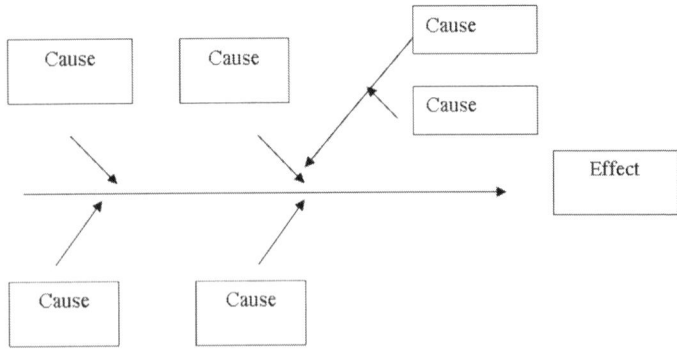

Figure 5.02

Audit

At an organizational level, managers expect see success in business terms – in increased profits Such evidence will keep their interest in the cultural changes involved in the 5S program. To paraphrase Lord Kelvin,

If you can not measure it, you can not improve it. This truth is essential to the understanding of the 5S program.

Once we have assigned job roles and integrated the 3S steps into the regular work pattern, then auditing, or evaluation of how well 3S conditions are being maintained, should be introduced to the program.

We are all familiar with measurements. Measurements have traditionally aided us in our daily lives. Without measuring what we are doing, we are unable to check whether the standards for the 5S program are being maintained.

Standardized Checklists

For the purpose of measurement, standardization-level checklists can be developed to record ratings for the five categories of 5S, while a general-level rating can be applied to the target areas. The ratings then show how well the area is doing in terms of 5S status. A scale of 1 to 5 is usually sufficient for this purpose, with a low rating (1 to 3) suggesting opportunities for improvement and a high rating (4 or 5) showing that the 5S program has been understood, implemented and sustained.

Auditing should be done at regular intervals and before and after 5S events such as big cleaning days or a 5-minute 5S (a common term in which the time can be changed to whatever is appropriate) to show progress.

Example Audit Checklist Figure 5.02 and Figure 5.03 of the audit check sheet being used in a retail garage

Standardization level Checklist		Dept:	Date			
		Assigned area:	Checked by:			
	Process - Checkpoint	Sort	Set in Order	Shine	Total	Previous total
1		1 2 3 4 5	1 2 3 4 5	1 2 3 4 5		
2		1 2 3 4 5	1 2 3 4 5	1 2 3 4 5		
3		1 2 3 4 5	1 2 3 4 5	1 2 3 4 5		
4		1 2 3 4 5	1 2 3 4 5	1 2 3 4 5		
5		1 2 3 4 5	1 2 3 4 5	1 2 3 4 5		
6		1 2 3 4 5	1 2 3 4 5	1 2 3 4 5		
6	Averages	1 2 3 4 5	1 2 3 4 5	1 2 3 4 5		

Figure 5.02

Standardization level Checklist			Dept: Service	Date 20. July. 2008		
			Assigned area:	Checked by:	Preecha	
	Process - Checkpoint	Sort	Set in Order	Shine	Total	Previous total
1	Tool room	1 2 ③ 4 5	1 2 3 ④ 5	1 ② 3 4 5	9	7
2	Technicians tool boxes	1 ② 3 4 5	1 2 ③ 4 5	① 2 3 4 5	6	5
3	Work bays	1 ② 3 4 5	1 ② 3 4 5	1 ② 3 4 5	6	5
4	Back Office	1 2 3 ④ 5	1 2 3 ④ 5	1 2 3 ④ 5	12	10
5	Customer Reception area	1 2 3 ④ 5	1 2 3 ④ 5	1 2 3 ④ 5	12	10
6	Customer Car Park	1 2 ③ 4 5	1 2 ③ 4 5	1 2 ③ 4 5	9	9
7	Averages	1 2 ③ 4 5 3	1 2 ③ 4 5 3.3	1 2 ③ 4 5 2.6	9	7.6

Figure 5.03

5S Audit Report Form

Once the audit has been completed, an audit report should be generated to cover all the findings, figure 5.04. A digital camera is an excellent tool for recording progress.

5S Audit Form	Area		Auditor Name		
Date					
No. Subject	Contents		Responsibility	Due Date	Status

Figure 5.04

70

Summary

Standardize is the result of properly maintaining the first three steps: Sort, Set in Order and Shine (3S).

Setting standards allows the organization's management to bring activities that would normally have varying results under control.

This stabilizes the daily management routine and reduces waste and loss. For any process to be strong, all steps within it must be equally strong (we will understand process management in chapter 8), but the secret to success is in making Sort, Set in Order and Shine (3S) unbreakable. t

Preventive measures for Sort mean that instead of allowing items to accumulate, we find ways to prevent their accumulation and prevent unnecessary items even entering the workplace in the first instance.

Set in Order breakdown can be prevented by using techniques such as 5W1H, mistake proofing (making it difficult or impossible to put things back in the wrong place) and even eliminating the need for the material, equipment or document.

Preventive Shine concerns preventing things from getting dirty. Its objective is to treat the cause of the contamination at its source, thus make Shine measures more effective.

(Continuous Sort) + (Continuous Set in Order) + (Continuous Shine) = Standardize

Points to Ponder

Take a few minutes to think over the chapter, then note down your answers to the following sample questions:

- ➢ What measures can you use to prevent the build-up of unneeded items in the workplace?
- ➢ Think of one Five-minute 5S activity that could be done each day to improve the work environment.
- ➢ Where could you use Visual 5S in your workplace?
- ➢ Using the 5W1H method, can you remove a constraint from a process and improve the efficiency of the process?

Chapter 6
S5 – Sustain
(Shitsuke)

- ➢ What Does Sustain Mean?
- ➢ Organizational Level
- ➢ Policy
- ➢ Process
- ➢ Job
- ➢ Importance of Sustain
- ➢ Implementation
- ➢ Summary
- ➢ Points to Ponder

What Does Sustain Mean?

In the context of the 5S program, "Sustain" means to make a habit of correctly maintaining 5S procedures. The previous chapters have taught us how to implement and standardize the procedures, but, without a discipline to follow, little benefit can be gained from the 5S program. Having taken the trouble to learn the tools and techniques of **Sort, Set in Order, Shine and Standardize**, you would think that **Sustain** should be easy. On the contrary, Sustain is a difficult element to achieve. Sustain, therefore, means discipline.

Organizational Level Policy

Senior management should focus on the organization level and assume the task of setting the mission and vision to direct the organization's growth. Every employee needs to know the principles and values of the organization – where it wants to go or how their behaviour is supposed to change to accomplish the organization's objectives.

A suitable environment should be created to encourage employees to change the culture and processes and support should be provided for employees to develop skills to optimise performance . This requires an efficient management system, such as TQM, to promote the development of a learning organization.

Executives and senior managers have the crucial role of encouraging continuous implementation and constant improvement so that 5S activities become the organization's norm and, ultimately, the corporate culture. This leads to a sustainable management system.

Process Level

Line management must emphasize processes. Sustain at the process level is mainly about the operation of systems that need to be followed and the achievement of the organization's policies and targets, by focusing on encouraging continuous implementation and improvement so that they become the norm ultimately embedded into the organization's culture. This leads to a sustainable management system.

Job

Line managers supervisors, and operational staff must focus on the job itself. The question is how to do this without causing waste or loss. In the hands of line managers and supervisors, 5S is a powerful tool for running the working body. The end result will be services and products tuned to customers' needs and demands. Managers should give regular training to operational staff to emphasize and support the discipline of **Sort, Set in Order, Shine and Standardize** – processes S1 to S4– so that they become the corporate culture .

Importance of Sustain

You commit to sustaining a particular course of action or program usually because the rewards for doing so are greater than the rewards for departing from the course. Viewed from another perspective, the consequences of not keeping to the program may well be greater than the consequences of keeping to the program. A personal fitness program is a good example; the rewards of going to the gym outweigh the rewards of not doing so.

The same principle applies to your implementation of 5S. Without your commitment to sustain **Sort, Set in Order, Shine and Standardize**, the benefits will soon fall away.

To illustrate this point, here are some of the things that happen when the commitment is not sustained. Figure 6.01 Boxes stored in a stairwell and figure 6.02 a workplace with nothing done to keep it clean

Figure 6.01

Figure 6.02

Implementation
Creating Conditions to Sustain Your Plan
Business Plans for 5S

Often managers do not plan key projects. Once the need for change has been recognized, there's a tendency to rush out to enthuse the organization about the need to change. This is also true for 5S. However, a lack of planning for 5S will result in a short-term program cynically referred to as the flavour of the month. Without a business plan, the final stage, Sustain, will be hard or impossible to achieve.

Every business plan should have, at the starting point, a vision of where it is heading and also a mission (statement) of what it intends to achieve.

What Do We Need in a 5S Plan?

Before starting to plan the 5S processes, we should have a clear idea about their objectives. Recall what has been learnt from the chapters on Sort, Set in Order, Shine and Standardize and then answer the following questions:

➢ What is 5S's intended outcome?
➢ What are our principal concerns about implementing 5S?
➢ What major problems do we hope to overcome with 5S?
➢ How will we convince everyone of the need to change?
➢ Who will manage 5S?
➢ How will we know that 5S is working?

Objectives of the Plan

The main impediments to a successful 5S program are the lack of:

➢ Vision and direction from senior management,
➢ Managers' understanding of what needs to happen and their role in 5S,
➢ Meaningful goals and measurements for success

The 5S business plan must therefore provide the framework for example a Balance Score Card as shown in figure 6.03 to address these areas:

- ➢ Specific goals and direction for results,
- ➢ Detailed activities to support systematic change,
- ➢ Details of resources required.

This means we must define in the plan the following elements:

- ➢ Vision and values of the organization,
- ➢ Management structure for 5S,
- ➢ Education and training for employees,
- ➢ Systems and tools for the 5S program,
- ➢ The main opportunities and priorities for improvement,
- ➢ Goals and criteria for success,
- ➢ Resources required.

Like all business plans, a plan for 5S should contain operational plans to implement it.

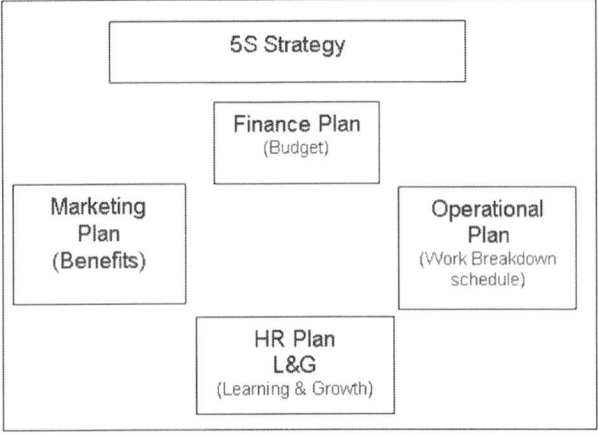

Figure 6.03

Adapted from A Balanced Scorecard

Internal Marketing Plan

To create the conditions that encourage the implementation of 5S, and the **Sustain** element in particular, we need to appreciate the decision-making system (DMS) that our staff will need to pass through:

Unaware   **Aware**   **Interest**
Desire   **Action**

The major role of managers and supervisors is to guide employees through the stages from unaware to action by creating the conditions that sustain 5S. Some of communication techniques are better at some tasks in this process than others.

Photo exhibitions and storyboards are powerful tools for promoting 5S; the saying, "A picture is worth a thousand words" is definitely true in the context of 5S. Photographs showing organizations before and after the implantation of 5S activities are an excellent method to communicate the status of 5S activities. Note the use of shadowing in the second picture, this helps to show if tools or equipment has not be returned. These photographs can also be used to demonstrate the standard that an organization would like to achieve.

Try not to judge the effectiveness of this kind of communication by an increase in improvements. It effects awareness and, to a lesser extent, attitude, but seldom does it directly shift people to the action stage

Before

After

Learning and Growth

A 5S program, like any other program, will involve a wide range of people with different skills and backgrounds. Care must be taken to select a mixture of employees with the skills and knowledge needed and to ensure that the team is balanced among the workforce, supervisors and management that will make up a successful 5S program team success.

Universal participation is a must. Some employees may ask to opt out of 5S tasks. The answer to this is no. Once 5S starts, everyone must follow the 5S work practices that have been established. Allowing employees to opt out of the process will make the **Sustain** stage impossible to achieve and can endanger the entire program.

Plan and implement **5S months** with activities such as seminars, workshops, contests and field trips. Sometimes the most powerful insights come from outside one's immediate environment because doing so can give one new perspectives.

Operational Plan

The Work Breakdown Structure (WBS) is the key document in a project such as 5S, as it forms the basis of much of the subsequent work in planning, setting budgets, controlling finances, defining the organization and assigning responsibilities.

The purpose behind the WBS is to structure the project into units of smaller, more manageable elements. This allows managers to assign specific authority and responsibilities. Each element should be independent and have as few interface dependencies with other elements as possible. The elements must be measurable so that progress can be demonstrated and monitored, preferably in a visual format.

By following a process of gradually dividing a piece of work into something more manageable, we will finally arrive at a piece of work that we can be plan and control.

Finance Plan (Budget)

Why budget? It is important to establish how much the 5S program will cost and how much the teams can spend on its implementating 5S. One method for allocating costs is to set a special budget account for the implementation phase. Then, once the 5S program is established, allocate the costs to the departments' cost centres. This will help ingrain the 5S activities into the day-to-day work and build accountability and responsibility.

Summary

Sustain means to make a habit of correctly maintaining 5S procedures. No matter how well the first four elements are implemented, the 5S system will soon stop without a strong commitment to sustain it.

Our commitment to sustain a particular course of action or program is usually maintained because the rewards for doing so are greater than the rewards for departing from it. This is true for any endeavour we undertake. In the same way, if the rewards for implementing Sort, Set in Order, Shine and Standardize for business and its staff are greater than for not implementing them, then this last element, Sustain, should be something you do naturally.

Sustain, unlike the other elements, cannot be implemented by a set of techniques, and it's not possible to measure it in a direct way. You can, however, create a suitable environment to encourage a change of culture and develop 5S skills so that commitment is optimized.

Executives and senior managers have the crucial role of encouraging continuous implementation and constant improvement of 5S activities, which will lead to a sustainable management system and demonstrate management's involvement and commitment.

Photo exhibitions and storyboards are powerful tools for promoting 5S. Photographs showing workplaces before and after 5S activities are an excellent method to communicate the status of 5S activities. They can also be used to demonstrate the standard that the organization would like to achieve. Other methods can include 5S department tours, field trips, seminars and 5S months.

How do you interpret 5S? How do you implement it? Do you see it simply as a way of removing waste from a process or as a way to control a process or encourage workers' learning and growth? It is all three. This needs to be acknowledged if 5S is to succeed; too narrow a view of 5S's purpose misses a lot.

Implementing superficial elements of 5S processes rather than understanding its place in a network of processes will cause you to struggle to achieve the benefits. A multiperspective approach can help keep the program from being superficial and is built into the learning process of team leaders. It is "management by means"[4] rather than "management by results" – by understanding the process and not being diverted from it by the natural variation common in all natural systems. Focus on the processes, and the results will follow.

Last, but not least, *universal participation is a must for success*.

Points to Ponder

> ➤ What ideas could you use to help make 5S work in your organization?

---

4   H. Thomas Johnson. Retzlaff Professor of Quality Management
School of Business Management Administration
Portland State University
Portland, Oregon, USA

➢ What would help sustain people's interest and motivation in and commitment to the 5S program?

Remember to write down your responses and ideas of how you can build them into your plan.

Chapter 7
5S Review

- ➤ Review
- ➤ Benefits of 5S
- ➤ How to Make 5S Work Effectively in Eight steps
- ➤ PDCA Continuous Improvement Cycle (Deming Cycle)
- ➤ Benchmarking
- ➤ Standards
- ➤ Summary
- ➤ Points to Ponder

Review

Over the previous six chapters, we have looked at how to implement the 5S "housekeeping program" and, by adopting **5S** at every level of staff, how it will strengthen the daily management of an organization. We have seen that it is an easy technique to apply and speaks the same language for the whole organization. Everyone from the top management to front-line staff can participate.

When we look at the lines of command for organizations figure 7.01 they can in must cases be be divided into three levels:

Organizational and Policy - Process – Job

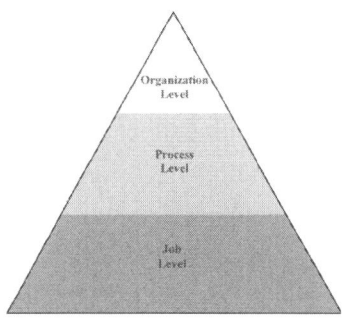

Figure 7.01

The 5S program can also be divided into these three levels:

You will recall that 5S is made up of the initials for five Japanese words Seiri, Seiton, Seisou, Seiketsu and Shitsuke, which translate to Sort (organize), Set in Order (tidy), Shine (clean), Standardize and Sustain (discipline). These 5Ss are the five keys to a total-quality environment.

In every company or organization, the pursuit of quality improvement must begin with the basics, and the 5S program – a campaign dedicated to organizing the workplace, keeping it neat and clean and maintaining the standard conditions and discipline needed to do a good job – fits this requirement. By adopting these 5Ss across a company or organization, you can yield tremendous results – the prevention of accidents, reduction of downtime, enhancement of operational control of processes and the creation of a healthier corporate climate.

According to Takashi Osada, author of *The Five S's*, we have practiced each of the 5Ss for a very long time, but we were not aware of them. In fact most of us will practice some form of Sort and Set in Order at home when we keep things like laundry baskets and waste bins in familiar places. If we look around a place, whether it is our home or our work area, we can always find room for improvement.

Benefits of 5S

> ➢ A clear workplace leads to high in productivity
> ➢ A clear workplace ensures high quality
> ➢ A clear workplace helps reduce cost
> ➢ A clear workplace ensures on-time delivery

- ➢ A clear workplace is safe for people to work.
- ➢ A clear workplace builds high morale.

How Do We Introduce 5S?

- ➢ Tidy up (active **5S**)
- ➢ Make **5S** a habit
- ➢ Take **5S** to a higher level (preventive **5S**)

Tidy Up (Active 5S)

Preparation: Recording the present situation

- ➢ Eliminate unnecessary items – **48-Hour Rule**
- ➢ Arrange storage places
- ➢ Consolidate daily cleaning procedures
- ➢ Maintain a spotless workplace
- ➢ Implement visual control in the workplace

Can you see the improvements?

Make 5S a Habit

Preparation: Photographing the new look of the workplace

- ➢ Control stock level
- ➢ Make it easy to use and return things such as tools and equipment
- ➢ Make clearing and checking a habit
- ➢ Maintain a spotless workplace
- ➢ Maintain standards throughout the organization

Has 5S become a habit?

Take 5S to a Higher Level (Preventive 5S)

Preparation: Evaluating the workplace where **5S** has become a habit

- ➢ Avoid adding unnecessary items
- ➢ Avoid disorganization
- ➢ Clean without getting dirty again
- ➢ Prevent degradation of the environment
- ➢ Systematize training

Auditing

- ➢ Organize a periodical audit and performance evaluation.
- ➢ Fix targets.
- ➢ Assign knowledgeable evaluators.
- ➢ Collect statistical data.
- ➢ Ensure continuous improvement.

Has your facility become a first class 5S operation?

Can 5S Be Carried Out in Any Environment?

The 5S program can be carried out in any environment. Factories, garages, hospitals, schools and retail shops have all applied these principles. Modern offices, for example, are easy places to implement them. But we can face problems as people have individuality and their own style of working. This can make Standardize difficult. Takashi Osada suggests "One" as the best campaign. Some examples of One are:

> ➢ One-day processing
> ➢ One file.
> ➢ One-hour meeting.
> ➢ One-page memo or email.
> ➢ One-minute phone call.
> ➢ One-copy filing

How to Make 5S Work Effectively in Eight Steps

Senior management support is vital, but whatever senior management accepts or wants will not necessarily be supported. We have to bring about conviction in the staff and clarify the concept by defining our mission, vision or goal for 5S and beyond by aiming for constant improvement. You cannot do this without maintaining a high level of motivation and satisfaction in the people that comprise your organization. Therefore, senior management should consider motivation and satisfaction part of the goal of professionalism and commitment. The following eight steps are designed to assist you in generating that motivation and making 5S work effectively.

Team Leadership

For senior management, this means providing strategic leadership which involves defining the overall vision and mission of an organization.

As with any improvement program, top management must drive implementation in order to create the environment needed for change and for long-term viability of the program. This can be done by choosing your team carefully, ensuring that they are from the right positions in addition to ensuring they have the right skills. The people you want to attract to lead a 5S project are those who come willingly. Once you have the right team in place, which should ideally be multilevel as well as multifunctional, it is essential to start in a positive manner. Encourage teamwork by inviting everyone to an informal meeting at the beginning to record the program's existence and clarify its purpose.

Infrastructure

Build the 5S project so that it fits within the existing organizational structures. For example, there are many similarities between the ISO 9001 requirements for quality management and 5S principles, and these should be integrated. This integration can be achieved by extending the ISO 9001 template to incorporate relevant 5S principles. By piggybacking on ISO 9001 quality management systems, 5S principles can be introduced more readily into organizations without the need for additional resources. These two sets of requirements and principles, when joined together, will help you move towards Total Quality Management (TQM).

Build 5S subcommittees within the project management group to deal with communication, training, project support and best practices – divide and conquer.

All levels of the organization must be involved, starting at the top. Eliminate organizational and physical barriers to teamwork. Eliminate performance ratings. Emphasize stability and constancy of effort – steady small gains rather than disruptive crash programs. Avoid the unsettling changes that result from not involving the whole team.

Communication

In chapter 6, we discussed the need to shift employees through the stages from unaware to action. This is a major task and challenge for managers and supervisors because it creates the conditions that sustain 5S. This can be accomplished by conducting short, focused and frequent communication sessions.

Consider delivering messages in several formats, such as group meetings, intranet, e-mails, newsletters, bulletins and posters. Some communication tools and techniques are better for some tasks in this process than others; for example, posters and notices can become wallpaper. There's an old saying "If you want to keep a secret, put it on the company's notice board". Also, delivering the same message in different formats can be confusing.

Ensure the Message Is Up to Date and Clear

There are often few opportunities for staff to keep management informed, yet those staff members who speak up are likely to understand the work problems better. Communication is a two-way process and requires two key changes in management behaviour to succeed:

> ➢ Learning to listen,
> ➢ Empowering staff to set the agenda for communication.

Train Teams

Within your 5S business plan, you should develop a plan to train everyone in the basic concepts of 5S. You will also need to supplement training with other information, such as the PDCA continuous improvement cycle (Deming cycle). We will return to this later, as it has an important role in the implementation of 5S. Other problem-solving techniques and root-cause analyses should also be used, such as cause-and-effect diagrams, which are used to investigate the cause of a problem and its effects, and fishbone diagrams (or Ishikawa diagrams). You will also need to consider how to measure your results; therefore, you will want to have people who understand statistical quality control. Considering the diagrams of warehousing picking errors that follow figure 7.02, you will see that the series data is not a straight line. Over time, all processes fluctuate up and down. This is natural variation in the work process, and it is not a cause for concern.

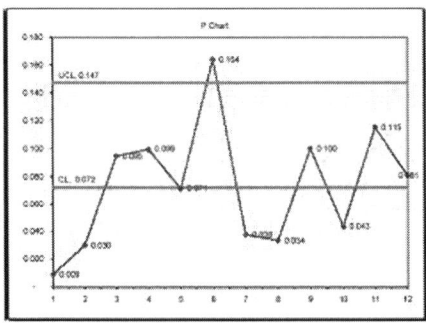

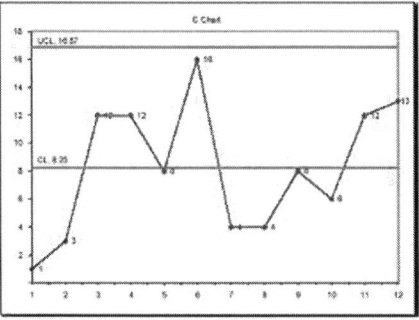

Figure 7.02

Attempting improvements could be expensive (economically wasteful) or make the process worse. What we are looking to identify are "special causes" – abnormal activity such as the one seen in the sixth month (June) on the P chart. When we asked the warehouse manager what the reasons behind this activity were, we found that he had to take on new picking staff because some employees had moved to higher jobs at the newly opened international airport, and the new staff required a period of training before the process could return to normal.

Creating a statistical process control (SPC) chart is a relatively simple task using Excel or a similar spreadsheet program. The following are the calculations for these charts, beginning with the data for picking errors:

Month	Number of errors	Variance
Feb	1	-
Mar	3	2
Apr	12	9
May	12	0
Jun	16	4
Jul	4	12
Aug	4	0
Sept	8	4
Oct	6	2
Nov	12	6
Dec	13	1
Total	91	40
Mean	8.27	4

1. Calculate the mean from the number of errors you have observed. In this instance it is 91/11 = 8.27).

2. Calculate the variation between two sets of errors, February and March for example (2 in this case), and so on down this list this should always be done as an absolute number so the results are positive.

3. Calculate the mean of the variation (for example, 40/10 = 4).

4. Multiply the mean of the variation by 2.66 (the constant used to calculate the control limits; for statisticians, it is the 3 sigma limit. In the example, 4 × 2.66 = 10.64).

5. Calculate the upper control limit by adding 10.64 to the mean of errors (8.27 + 10.64 = 18.91).

6. Calculate the lower control limit (LCL) by subtracting 10.64 from the average (8.27 − 10.64 = 2.27). Since this is a negative number, there is no LCL. The objective is to have zero errors.

Variation exists in everything. This is important to understand when using SPC charts. However, whether we're manufacturing products or performing services for customers, who demand high quality and consistency in our goods and services, variation can become a big problem.

Too much variation leads to rework, scrap, or customer problems. In theory, a perfect process would be one with no variation, but perfect processes don't exist. Just think of a flat line on an EKG; for a nurse, this would be a worrisome sign, as it signals no heart activity. But as the variation in our processes is reduced, the output of our processes is improved. That's our goal with SPC – to reduce the variation in our processes and then monitor the process to make sure that the variation doesn't increase.

When monitoring processes, be on the lookout for abnormal events that do not occur randomly. SPC control chart data can have many shapes that show some possible outcomes, as the following set of charts demonstrate:

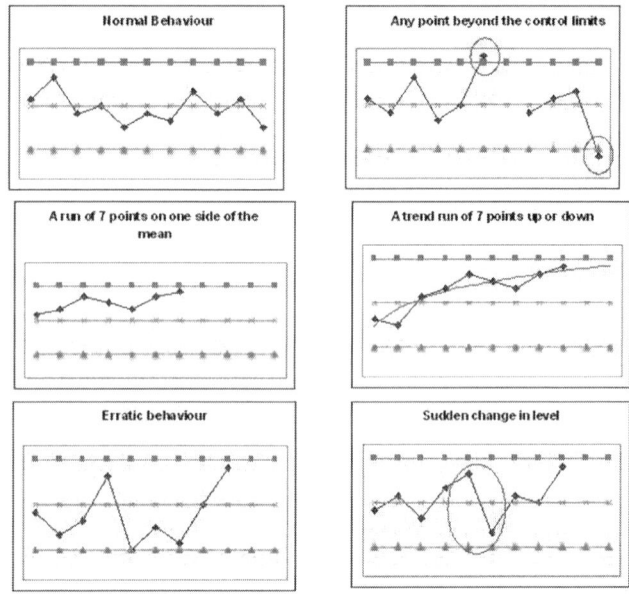

Even if the chart does not show some special cause for analysis, the data should still show examples of every fifteen observations to ensure that the process is not slowly drifting away from normal behaviour.

An SPC chart is an excellent way to demonstrate reduced adverse outcomes. By making these reductions visible, SPC charts make employees aware of their performance. When implementing, it is worth it to go a step further and bring together the SPC chart, analysis sheet, action plan, and implementation record into one measurement control chart.

The objective is to ensure performance is measured, causes of under performance are identified, actions to rectify the problem are planned and implemented then the performance is measured again, this closes the loop and makes sure performance is regularly reviewed and acted upon.

There are a few mistakes to avoid:

- ➢ Changing the process to accommodate the special cause. This usually adds cost and bureaucracy.
- ➢ Blaming individuals. Not only does everyone make mistakes, but also, chances are that the problem would have occurred regardless of who caused it.
- ➢ Pressuring workers to simply do better. People can only do as well as the system allows them to do.

Other skills that you need to provide training for are leadership, communication and presentation skills, and project management.

5S Pilot, Demonstration Projects

Pilot projects are usually the first projects to demonstrate the principles and approaches that an organization hopes to adopt later on a larger scale. Pilot projects are a vehicle for improving quality and promoting innovation. They allow problems to be solved before a new system is introduced more widely and give you the opportunity to demonstrate the feasibility of the 5S system. But they tend to have limited impact if they are not accompanied by explicit strategies for transferring learning. For this reason, they are more transitional efforts than end points, and they involve considerable amounts of learning by doing. Pilot projects implicitly establish policy guidelines and rules. Managers must therefore be sensitive to the precedents that are being set and send strong signals if they expect to establish new ways of working.

You may wish to test your new-found knowledge by running a trial before rolling out 5S company-wide. This can be best done by selecting areas that most need the 5Ss and have the highest visibility as pilot projects. What is learnt in these pilot areas can then be used to develop a full roll-out plan.

Pilot projects, however, often cause severe tests of commitment from staff, who wish to see whether the rules have, in fact, changed.

Establish Best Practices

Creating and using a best-practices database can enhance the success of **5S** by providing the means to share successes throughout an organization.

The ideal of a best practice is not new. Frederick Taylor stated one hundred years ago: "Among the various methods used in each element of a trade, there is always one that is quicker and better than the rest". This view was later called the "one best way". And is also associated with the **Motion Study** work of **Frank** and **Lillian** Gilbreth [5]

---

5 One best way is a forerunner to the development of continuous quality improvement.

While Frank and Lillian Gilbreths' work is sometimes associated with the work of Frederick Winslow Taylor, there was substantial

Best practice is a philosophy based on continuous learning and improvement. In recent years, it has become associated with the Japanese word "kaizen" and the term "kaizen events" that have been imported into Western organizational language to emphasize the importance of efforts to constantly improve. Benefits of best practice often include quality results and consistency when the process is followed.

The ideal of best practices does not tie you to one inflexible, unchanging practice. Instead, a best-practices approach is based on continuous learning and improvement and relies on three themes that resonate with successful benchmarking:

> ➢ Transfer is a people-to-people process – meaningful relationships precede sharing and transfer.
> ➢ Learning and transfer is an interactive process that does not rest in a static body of knowledge. Employees are inventing, improvising and learning something new every day.
> ➢ Best practices do not have one template or form for everyone to follow.

---

philosophical difference between the men. Taylor was primarily concerned with reducing the time of processes (the "one best way"). The Gilbreths sought to make processes more efficient by reducing the motions involved, and they saw their approach as more concerned with workers' welfare than Taylor's approach, which workers often perceived as primarily concerned with profit.

Benchmarking stems from a personal and organizational willingness to learn. A vibrant sense of curiosity and a deep respect and desire for learning are the keys to success.

Full Roll-Out Plan

An effective plan maps out your 5S project, detailing what needs to be done. Prepare your plan well, and it will guide you to success; that's your 5S business plan.

After the pilot projects are complete and before you set out on an organization-wide program, step back and evaluate what you have learnt from and achieved with members of the pilot projects. Take those ideas from them to include in the roll-out plan that will strengthen the 5S process and also add them into the best-practices database.

As part of the planning process of 5S, value stream mapping (VSM) can be used. This process documents the current and future states of information and material from customer to supplier. It looks at a specific set of actions – those that add value and those that do not – by eliminating actions that create no significant gains in value improvements can be made in productivity which will have a dramatic effect on your company's bottom line.

Continuous Evaluation and Adjustment

As with any process, as you learn lessons, also make improvements to your 5S effort, modify and strengthen the infrastructure, and develop improved methods. Measure and communicate progress at the same time. Challenge work areas to constantly improve.

Using kaizen events can be useful during in-context evaluation and adjustment. Kaizen is a workplace quality strategy and is often associated with the Toyota Production System that is built upon the learning-by-doing concept and related to various quality-control systems, including those developed by W. Edwards Deming.

Kaizen (改善), is simply Japanese for improvement. In the context of 5S it refers to implementing continuous improvement and emphasizes the learning-by-doing aspect. This philosophy differs from the command-and-control improvement programs of the mid-twentieth century in that it involves continually making small improvements.

PDCA Continuous Improvement Cycle (Deming Cycle)

Where does quality begin? Where and how do we improve it? Every day we struggle with the question, "how do we make things better?" The following two diagrams figures 7.03 and 7.04 reveal that the answer does not lie with our Quality Management System (QMS).

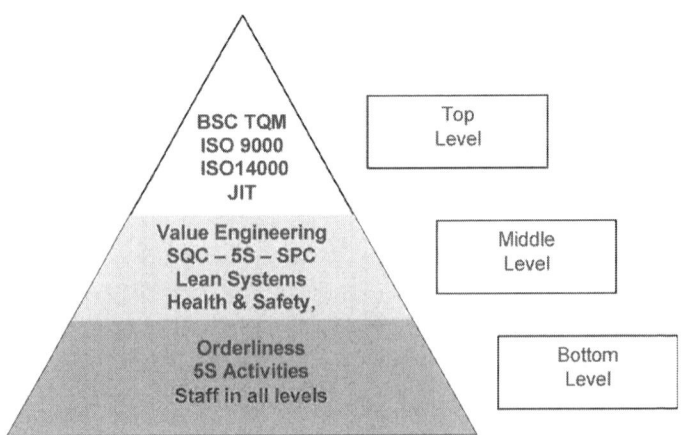

Figure 7.03

Optimization of organizational efficiency

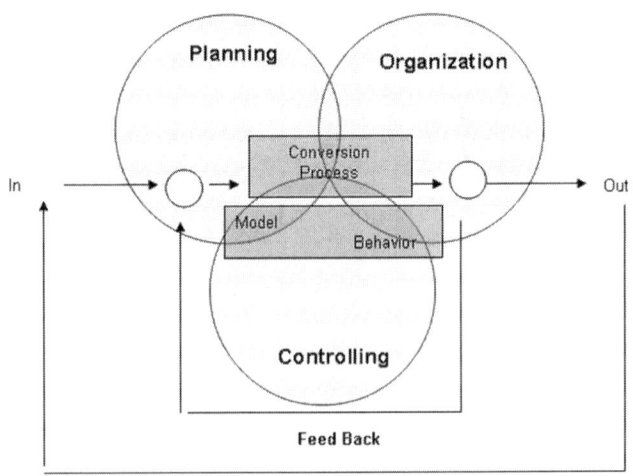

Figure 7.04

General model for managing operations

108

In fact, the continuous improvement cycle begins outside these models. It begins, W. Edwards Deming's view, with a "system of profound knowledge" .(The W Edwards Deming Institute) consisting of:

> ➢ Appreciation for a system,
> ➢ Knowledge about variation,
> ➢ Theory of knowledge,
> ➢ Psychology.

The various segments of the system of profound knowledge cannot be separated. Instead, they interact with each other. Thus, the knowledge of psychology is incomplete without the knowledge of variation, for example.

Managers need to understand that all people are different and that anyone's performance is mainly governed by the system they   work in.

The same can be said of a QMS, of which 5S is a cornerstone. Any QMS is a collection of subsystem processes, knowledge, variation and people who operate the system. All are interrelated. If we accept W Edwards Deming and Walter A. Shewhart's (actual creator of the PDCA cycle figure 7.05) view, then the enemy of improvement is variation. Our QMS has more to do with our organizational culture than any particular standard. Therefore, your QMS is whatever your organization decides it is with all its defined or undefined processes, people, variations and knowledge inherent in the system.

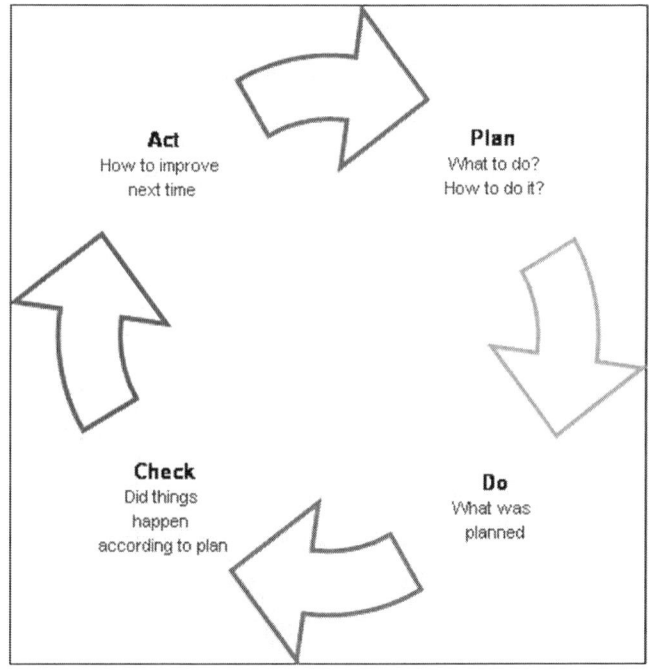

Figure 7.05
Plan – Do – Check – Act Cycle

The PDCA cycle is a four-step model for carrying out change. Just like a circle, there is no end. Therefore, PDCA process should be continuous.

When to Use PDCA?

➢ Making continuous improvements

➢ Starting an improvement project (such as 5S)

➢ Developing new processes, products or services

➢ Planning data collection and analysis

➢ Implementing any changes

Procedure

➤ **Plan** – Recognize the opportunity for change, and then establish the objectives and processes necessary to deliver results.

➤ **Do** – Implement the processes.

➤ **Check** – Monitor and measure the processes against the policies and objectives, and identify what you have learnt.

➤ **Act** – Take action based on what you learnt in the check step. If the changes did not work, then go through the cycle again with a different plan. If they were successful, fully incorporate them into your system and use what you learnt to plan new improvements. The results of one cycle become the check phase of the next cycle, and Act begins a new cycle. This is also called rolling the Deming wheel.

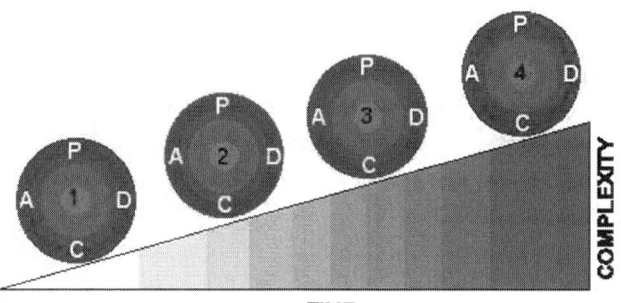

Figure 7.06

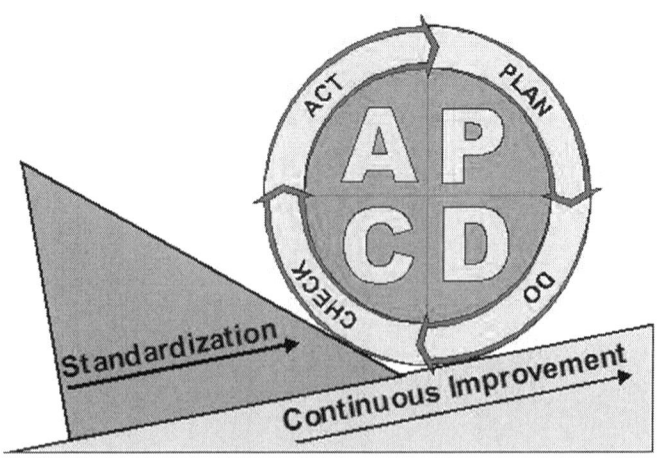

Figure 7.07

The previous two diagrams figure 7.06 and 7.07 demonstrate clearly what we are aiming to achieve: improving our organization over time and then incorporating those improvements as standards.

Benchmarking

What Is Benchmarking?

In its simplest sense, "benchmarking" is comparing one thing against another. Benchmarking is something we have all done at some point in our lives – for example, in sports, when we competed against others or even against ourselves (did you set a personal best time in at a swimming competition?).

Benchmarking is a process of investigation and learning that enables us to compare and perhaps improve our own activities.

Why Benchmark?

Not all learning comes from reflection and self-analysis. Sometimes the most powerful insights come from looking at other operations. Often organizations in completely different fields of business can prove fertile sources of ideas and catalysts for creative thinking.

The greatest benefits come from the study of practice benchmarks – the way work gets done – rather than those for results, and from involving line managers at the process and work levels.

How to Benchmark?

Without understanding how the 5S processes work within our own organization, our ability to apply the results of benchmarking will be limited.

Therefore, a thorough understanding of our own operating processes and an appreciation of their strengths and weaknesses is needed. A SWOT analysis is a useful tool for this purpose.

Strengths	Weakness
Opportunities	Threats

There are many views on the following steps in bench-marking. Outlined below is a simplified model based on King, Moran and Niall's "Benchmarking: An Operational Necessity,

> ➤ Analyze your process or processes and decide what to benchmark
> ➤ Define and measure the selected process and develop a benchmark plan
> ➤ Select a benchmarking partner and agree to the parameters
> ➤ Carry out the benchmarking
> ➤ Analyze results and apply them to your process
> ➤ Establish new metrics

These steps were first published in the Telecommunication Journal of Australian. Vol. 42 No. 3 1992. And in essence benchmarking can be viewed as the PDCA cycle in action.

Standards

It would be hard to overestimate the importance of standards in our lives. Standards in the context of **5S** are not moral or ethical standards but the standards of languages, symbols, conventions, scales. These shared values are the building blocks of civilization, for without our ability to develop and accept standards, we could never have developed such a complex society.

Why Use Standards?

➢ Standards enable us to communicate; every language is a set of shared standards.

➢ Standards drive learning;

➢ Shared standards make skills transferable.

➢ Standards make comparison possible. Around 1494, Luca Pacioli codified the double-entry bookkeeping that we still use to this day. This is a clear example of standardization.

➢ Standards foster creativity. Think for a moment about music. On the surface, it would seem that the chromatic scale would be limiting, but over the past two centuries, it has given us some great music precisely because composers had a standard upon which to express the unique music in their heads.

Standards are the code in which discovery is written and the means by which we collaborate. If we want to build a cooperative, creative organization, then we have to ensure we use the relevant codes.

Summary

The key to success is not only to implement consistent methods to improve efficiency, but also to make the 5S practices a vital part of an organization's culture.

Improvements based on industrial engineering disciplines can be applied to offices, retail outlets and repair workshops equally as well as manufacturing operations. All organizations will benefit from these improvements.

One of 5S's key benefits is that it sets a foundation for continuous improvement.

Organizations that use 5S programs will improve profit by instilling a culture of quality and safety in the workplace. The 5S method does this by reducing waste of all types – time, material, supplies and maintenance – and simplifies the working environment, thus raising morale and improving efficiency and productivity.

Resistance to Change

Any organization introducing a change strategy such as 5S is likely to encounter resistance that needs to be addressed through motivation and training. One of the benefits of the program is that resistances, including the fortress management style, that may be present in an organization can be overcome through motivation and training.

Managing change is political. There is nothing more difficult than to plan, nothing more doubtful of success than a new system. The old school prefer the preservation of old ways, and those who can gain from new ways are likely to give them only lukewarm support.

Applying 5S to your organization will be hard but worthwhile work. To achieve success, you need to break down barriers between departments.

Management should learn their responsibilities, take on leadership, and improve constantly.

Instigate a Program of Education and Self-Improvement

As individuals or organizations get more experienced at a task, they usually become more efficient at it, following a typical learning curve. Shared experience affects the experience curve. The effects are reinforced when two or more groups share a common activity or resource. Any efficiency learnt from one group can be applied to the others, creating team learning.

Team learning happens when a group of people working on something together experience that rare feeling of synergy and productiveness that happens when you're "in the groove." When a team is truly learning, the group as a whole becomes much more than just the sum of its parts. Practicing this discipline involves an array of different kinds of conversations and a remarkable degree of honesty and mutual respect during participation. The object is to create a shared vision for your organization. It will emerge when everyone in the organization understands what the organization is trying to do why it is trying to do it, and when everyone is genuinely committed to achieving that vision, and clearly grasping how his or her role in can contribute to making the vision real. Practicing this discipline involves knowing how all the parts of the organization work together and being clear about how your own personal goals align with those of your organization.

As we have discussed, 5S is a visual program that relies on visual communication, such as pictures, to transfer knowledge within a group. It aims to improve the transfer of knowledge beyond just the transfer of facts. The aim is the transfer of insight, experiences, attitudes, values, expectations, perspectives, opinions and predictions by using various complementary visual techniques.

Points to Ponder

The visual control system is a whole philosophy, not just a process for pinning charts and graphs on the notice broad

The following characteristics can be expected to be found when an effective Visual control system is being used.

- ➤ **Self-explaining** – Anyone walking through the area can tell how the workplace "works" through symbols, pictures, and other visual devices.
- ➤ **Self-ordering** – Colour coding, floor markings, tool outlines, bin markings make it clear where things belong and when they need to be replenished.
- ➤ **Self-regulating** – the information needed to monitor and regulate tasks, production rates, quality, and other factors are prominently displayed.

Self-improving – Visual control system is a process of continual improvement and employee involvement

- ➤ This chapter has reflected on the 5S program and how to implement it. Give yourself a little time to consider what benefits your organization needs and how you can achieve them by using the 5S program. Write down your thoughts.

The last chapter, on management by means, will help put the 5S program into context with an overall system of continuous improvement.

Chapter 8
Process Management Approach
and
Management by Means

- ➢ Introduction
- ➢ Management by Means (MBM)
- ➢ Values and Principles
- ➢ Processes
- ➢ Requirement based output
- ➢ Correct from me
- ➢ Continuous Improvement
- ➢ Summary

Introduction

Like any other work, 5S is made up of sets of process-es, which means we need to consider process manage-ment in a little more depth as all work requires analysis, planning and doing. While each of these elements is a separate part of the same job, they are not separate jobs divorced from one another. For work (a process) to be done effectively, all three elements must be present.

Management by Means (MBM)

There are many tonics for organizational develop-ment, such as Six Sigma, Seven Habits, Six Hats, Mind Maps; the list sometimes seems endless. In fact, it's im-possible to study or implement them all. Using all tonics would be like taking a wide range of medicines to cover all possibilities when you feel unwell; it's fixing the problem by using a wandering–sheep-syndrome method.

Figure 8.01 illustrates a technician applying the wan-dering–sheep-syndrome method to fixing a problem on a car.

Figure 8.01
Wandering sheep syndrome

But you will need to use more than one approach to manage an organization. The 5S program we have been discussing throughout this book is just one component of the jigsaw on the way towards demonstrating quality in an organization.

Continuous improvement and MBM – the concepts we are about to discuss – are about understanding processes and not getting diverted from the task by the natural variation common to all systems

The results will follow' a kaizen process of continuous improvement, **an** incremental, but constant, process also known as evolutionary development.

While 5S implemented on its own will benefit an organization, it should be seen in the wider context of organizational management as illustrated in the diagram that follows. Figure 8.02 diagram was first introduced in chapter 7 as a model for producing maximum organizational efficiency.) The theory and concepts of MBM can be applied to all the steps in the diagram.

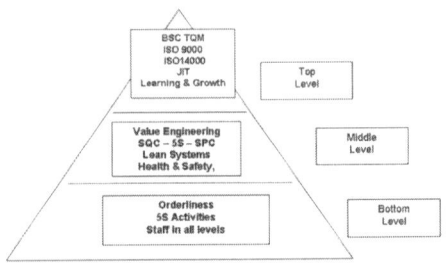

Figure 8.02
Optimization of organizational efficiency

Before discussing MBM, it is worth reflecting on the structure of organizations by reviewing their main components – starting with a set of shared values and supporting principles as the foundation and concluding with a vision for the future that recognizes the potential of an organization figure 8.03

McKinsey's "7S Framework figure 8.04 which was developed by Tom Peter and Robert Waterman while working at the consulting firm McKinsey & Company suggests that there are seven components which must fit together if strategies are to be effective:

> **Strategy** – A clear set of actions aimed at achieving sustainable improvements,
> **Structure** – Clear organization of responsibilities, represented by the organizational chart that shows who reports to whom and how the tasks are allocated,
> **Style** – Tangible evidence of what is considered important; not what management says, but the way it behaves,
> **Staff** – The people in the organization. These people think more in terms of the group mix and collective skills than individual personnel,
> **Shared values** – The principles that guide us and cement the organization together,
> **Skills** – These are derived from the other six elements and are the capabilities the organization has, not the people in it.

Figure 8.03

Organizational model

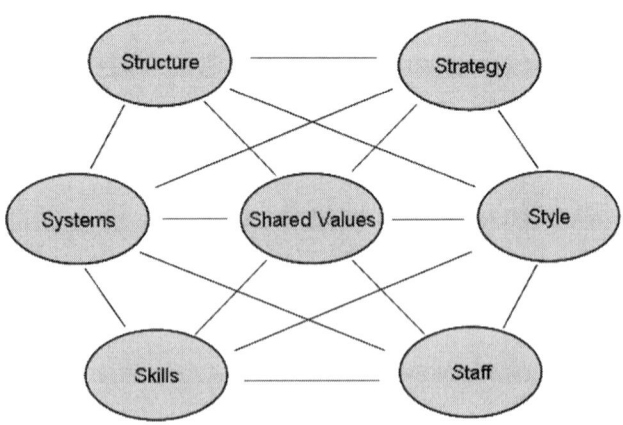

Figure 8.04

McKinsey's 7S Framework

The lack of a declared purpose, vision and shared values is one of the most common reasons why organizations fail to deliver their very best.

Values are those timeless principles that guide us. They define what we stand for. Each organization must discover its core values. Values and principles are the cement holding organizations to together.

H. Thomas Johnson's Management by Means [6]

Management by Means (MBM) theory features the concept that our work is motivated by common values and principles and not just by preconceived targets. This concept is in contrast with "Management by Results" (MBR), which is based on achieving explicit targets. With MBR, you must consider whether the measurement really delivers what you intended as a result and what behaviour the measure encourages, and whether the behaviour is desirable to obtain the results – whether the means justifies the end result. MBM is process orientated. In other words, by focusing on the process and ensuring it is correct, then the result will follow.

---

6 School of Business Management Administration
Portland State University
 Management by Means
H. Thomas Johnson, Retzlaff Professor of Quality Management

Portland, Oregon, USA

Note that MBR is also often confused with Peter F Drucker's concept of Management by Objectives (MBO) but should not be. A close study of Drucker's work will dispel this notion.

Johnson's MBM reflects the views of W Edwards Deming and Gregory Bateson in two important aspects: both theories see organizations and society as natural systems from a relational and evolutionary perspective. Deming and Bateson also have in their works the recurring theme of the detrimental impact on the quality of life if quantitative goals are pursued without regard to their effects on people's relationships with each other – a nature which leads to the concept that an organization is a living system and as such is governed by self-organization, interdependence and diversity.

Self-organization is the capacity for living systems [7] to define and sustain themselves with a unique identity even while they continually adjust and adapt to feedback from their environment. Self-organization implies ability for continuous growth if it a living system is not restrained by its interdependence on other systems. Every system in nature is interconnected to every other system, and

---

7   Living systems are open, self-organizing systems that have the special characteristics of life and interact with their environment. These can range from a single cell to a complex institution, such as a business or university. All rely on sets of subsystems (processes) to function, arranged as inputs, throughputs and outputs. James Grier Miller, Living Systems Theory *(McGraw-Hill, 1978).* International Society for the Systems Sciences (ISSS)

any system seeking to utilize all the energy or resources for its own purposes is bound to be challenged by other systems. The consequence of these interactions between self-organizing systems is a continuous stream of new things, or in the case of humans, new thinking. This is diversity.

Bateson interpreted self – organizing systems as working together to sustain the existence of an evolving ecosystem. This approach has its roots in Alfred Russell Wallace's work. Wallace saw that the job of evolution was to maintain the constancy of something in his case, the entire ecosystem made up of all species and their environment – a process rather like the cruise control system or constant velocity transmission (CVT) on a motor car. We can also think of it in terms our bodies' ability to adapt to changes in the outside temperature, at least within a limited range. By shivering or perspiring, our body temperature remains more or less constant because we vary internal conditions in response to those changes in outside temperature.

Johnson's theory applies this principle to organizations. In doing so, the organization is then being "managed by means". To practice MBM, managers must gain knowledge of the many elements inherent in the processes and the work that is being done. They can pay attention to the people who perform that work by giving them the opportunity to make decisions, collaborate and recognize and solve problems and also develop new approaches to tasks which will bring about innovation that will benefit the organization and society.

Values and Principles

These are the opening steps in establishing the concept of MBM. They are the essence of what we do and the guide of how we work. These are illustrated in the process management model of the author's organization, with values forming the base figure 8.05

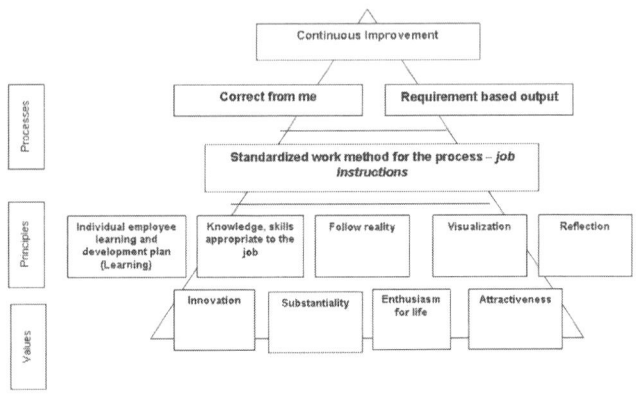

Figure 8.05
Process Management Model

Values

First, Innovation is the search for solutions to customers' or society's needs that will bring benefits to society. Then, Substantiality plays a crucial part when Services to the customer is solidly built on skills needed to satisfy the customer's requirements

These two values lay the foundation for sustainable development, which includes eliminating waste from processes. Enthusiasm for life is an expression of friendliness and integration, showing respect to others and for nature, treating everyone as intelligent friends while at the same time maintaining a sense of balance and humour. This enables us to look at the things we do as being not just for a result or profit but also as being a means of improvement for the betterment of society and the environment.

Attractiveness: To succeed the brand(organization) needs to show passion for providing exceptional products and services, allowing society to have a sense of optimism about the organization and share in those values, thus being attracted towards the organization.

Principles

Principles, guide the activities of the management processes, continuous improvement requires a commitment to the first principle, learning. The statement, "continuous improvement requires a commitment to learning" may seem obvious, but quite often it takes a back seat, with support for learning objectives reduced to focus from long-term to more immediate business issues if the economy is suffering a downturn. But without learning, organizations and people would simply repent old practices repartition, and there would be no development. Continuous improvement and effective learning not only requires a commitment to learning, but also motivation – wanting to learn.

The second principle, **knowledge**, concerns having the right knowledge and skills to perform the job. This means that it is important not only to provide training **[8]** but also to retrain and upgrade skills and information as well; however, knowledge is pointless unless you do something with it, such as solve problems or try something new. Because only by solving problems and trying something new are we able to gain understanding and knowledge (see 'knowledge and understanding diagram') figure 8.06. If we apply this learning often enough, we will have the added value of experience. It is very difficult to become knowledgeable in a passive way; therefore, for the learning to be effective, you need to do something. Only by applying new knowledge to the way work gets done would the potential for improvement exist.

---

8   "It's all to do with the training: you can do a lot if you're properly trained." – Elizabeth II, Queen of Great Britain and Northern Ireland (b. 1926). Television documentary, BBC1, 6 Feb. 1992. In early 1945 the Princess was made a Subaltern in the Auxiliary Territorial Service (ATS). By the end of the war she had reached the rank of Junior Commander, having completed her course at No. 1 Mechanical Training Centre of the ATS and passed out as a fully qualified driver
http://www.royal.gov.uk/HMTheQueen/Publiclife/EarlyPublicLife/Earlypubliclife.aspx

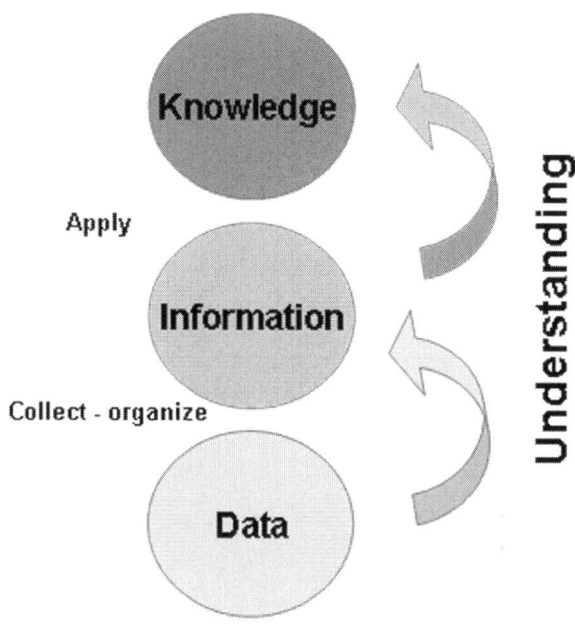

Figure 8.06

Knowledge–Understanding system
Knowledge is about perceiving, storing and retrieving data,
while Understanding involves reasoning skills.

The third principle, **follow reality**, simply means that what we are doing and learning is relevant to our needs today and tomorrow, and not in the past. The fourth principle, **visualization** aim, is the transfer of knowledge using various complementary visual techniques to improve the transfer of knowledge beyond just the transfer of facts. It's not just about making documents; it's about communicating – making information easy to find and to understand.

The illustration of the core process figure 8.07 for a vehicle service in a retail garage provides a clear example of visualization of the steps that need to be performed in the process of servicing the car and taking care of the customer's needs. It is worth noting that in the example, the number of processes with the external customers is greater than those with internal customers.

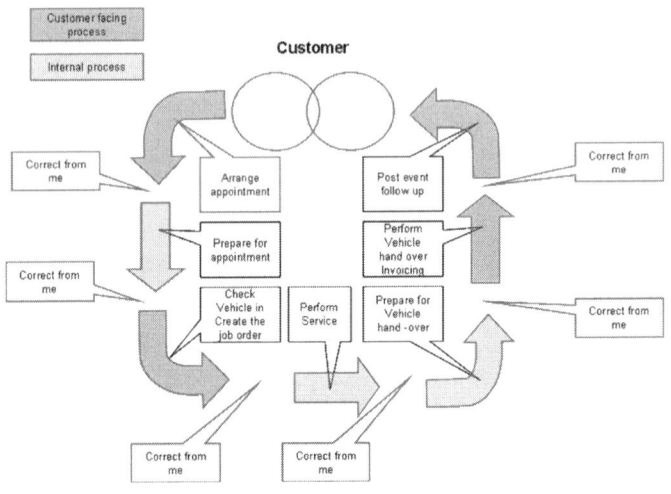

Figure 8.07
Service Core Process

Reflection, or "abstract conceptualism", is the central point that gives us the opportunity to draw conclusions about our practices. To learn and improve, we need to reflect– it is the process of reviewing and making sense of what has happened, then forming new ideas about the way of doing things in the future. But reflection by itself is not enough to promote learning. Unless we act on those reflections, we will not develop. Learning from our experience allows us to plan what to do next by bringing together theories and analyses of past actions, enabling us to come to conclusions about our practices.

Reflection needs to be a continuous process for any individual or organization. David. A. Kolb's Experimentation Learning Theory (ELT) [9] form the basis for a learning organization. In the illustration, ELT is depicted as a spiral rather than a cycle see figure 8.08.

---

9 Kolb's model gives us a routine and a process which we can use to continuously develop our working practices. The model suggests that it is insufficient just to experience something to learn; reflection is also necessary to enable the formation of concepts which can be applied to new situations by linking the theory and action by a process of planning, acting out, reflecting and relating back to the theoretical elements of our work.
Knowledge and understanding have improved beyond your original starting point. Nevertheless this new understanding represents a new starting point in your quest for even deeper understanding.

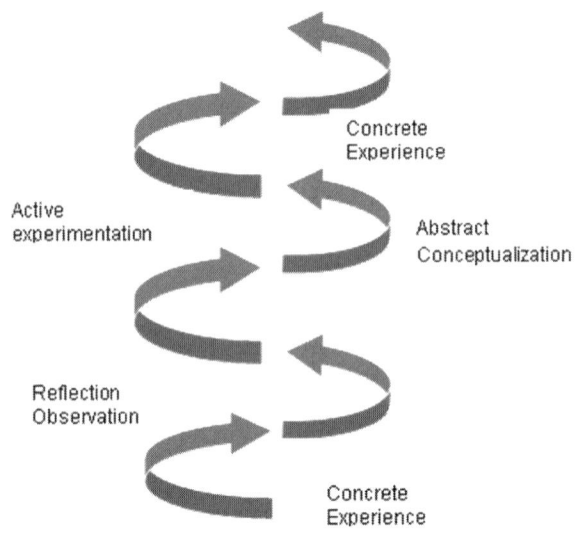

Figure 8.08

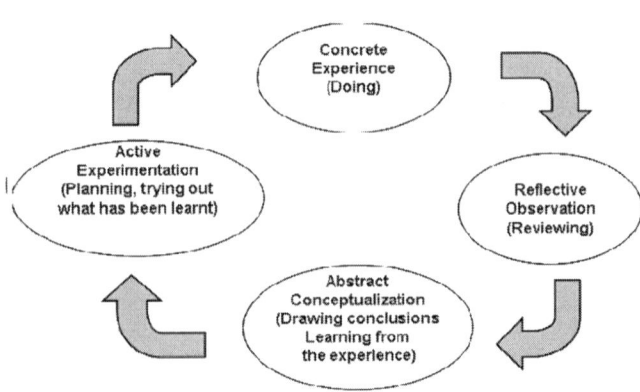

Effective learning comes from wanting to learn. It requires the desire to understand the process for finding appropriate solutions, not just the solutions themselves because in organizations, there is no perfect single solution. This means we need to be prepared to accept mistakes and gain insights from them. This is a key concept behind the "Correct from Me" stage, which we will look at a little later.

Processes

The principles in the above model are the working practices starting with standardized processes. These describe the preferred method for performing work in a consistent way, as in the core processes of servicing a car or planning a business. Most work take place without any particular disruptions or causes for concern, in effect, it results from normal behaviour. Knowing what is normal lets us to detect when something has deviated from the norm; knowing this is the best way to know what to improve.

Requirement based output

"Requirement-based output" is the work required to fulfil a need, and we should work towards that need. In doing so, we will avoid waste, which is quite often hidden in day-to-day operations and costly, often as much as 30 per cent of operating costs in service industries. It is worth pointing out that it is often quite simple to try to improve efficiency. But if the increased performance or process does not fulfil a need, then it's waste. For the system to work, see the Service Core Process illustration. We should next consider the principle of "correct from me", at each stage of the process, for this to work think of these three rules for error proofing

> ➤ Don't accept a defect from a supplier
> ➤ Don't make a defect yourself
> ➤ Don't pass on a defect to the customer

Correct from me

The "correct from me" principle says that everyone must pay attention to what they do. If everyone does this, then the system will work. Understand this principle and also bear in mind that all work is a process with the main system often made up of many other smaller processes (see the high-level to task-level flowchart). In linking together processes both internal and external to the organization and keeping these links intact lies the secret to quality work. See figure 8.09

Correct from me does not imply that we are not allowed to do things wrong and that everything must always be right the first time (with zero defects). What it is telling us is that we need to correct the things we find wrong and then look at how to improve. We should not put off trying new ideas for fear that we may go wrong. Rather, we should enjoy correcting things we find abnormal.

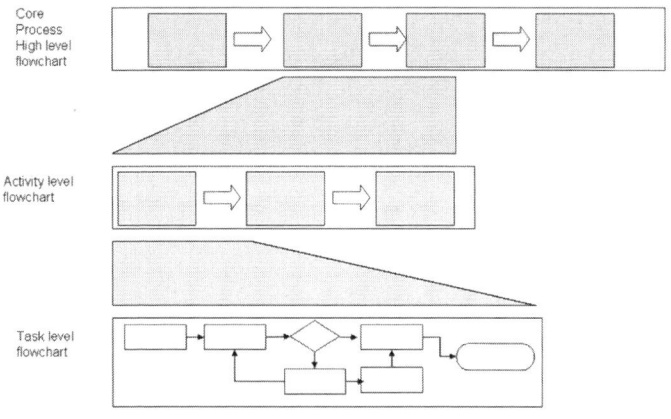

Figure 8.09
High-level to task-level flowchart

Continuous Improvement

The final element of the system, "continuous improvement",[10] emphasizes that no improvement is too small to ignore. Instead, small, incremental improvements which do not move too far away from the normal experience are the best means to achieve your goal. You will achieve the biggest gains from improving the way the system works and through cooperation among departments (an end to fortress management). These improvements in performance come from improvements in the processes, which are usually more sustainable than motivational programs.[11] While motivated personnel are important,

10   "A successful individual typically sets his next goal somewhat but not too much above his last achievement. In this way he steadily raises his level of aspiration." —Kurt Lewin

11   "balanced score card," Dr. Mike Bourne and Pippa Bourne. *Theory P* (Hodder Arnold, 2007), 150.

they often gain very little from being motivated to work a bit harder and make fewer mistakes, as ultimately all motivation comes from within ourselves. Motivation is a process of our personal attitudes towards the tasks we are engaged in. The illustration that follows demonstrates the components involved in the continuous-improvement process that affect our attitudes and motivation.

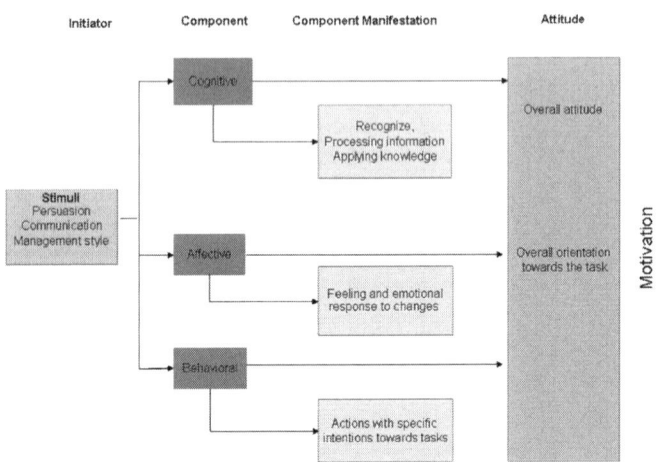

Summary

We began by highlighting an array of management tonics that can be used for organizational development. There is a danger of overdosing on them, which can result in fixing problems or trying to improve processes by using a wandering-sheep-syndrome method of management. This comes from a lack of clear understanding of the causes of problems or of the requirements and needs of the system.

To overcome this style of management, an organization can use the model of management by means (MBM), which emphasizes the need for organizations to have a declared purpose and vision and to share values. These are the timeless principles that guide us and define for us what we stand for and why. Each organization must discover the core values that are important to it and to the people within it. Our work is driven by these common values and principles, and not just by preconceived targets, when we use MBM.

We then progressed up the model to principles of individual learning and learning organization and their underpinning theory and process, the learning cycle. We emphasized:

> - Developing the right skills and knowledge to address today's and the future's needs,
> - Realizing the importance of communication,
> - Reflecting on what our processes produce and how to improve them.

With our understanding of the role that standardized processes play in producing quality work, we should also understand that the work itself must fulfil a need; otherwise, the work generates waste. This is the core of the idea of requirement-based output.

We must pay attention to what we are doing. If everyone does this, then the system will work. But when we observe something going wrong, we should put it right straightaway. Sustainable improvements come ultimately from what we do; therefore, no improvement is too small to ignore. And by adopting a process-management approach we will achieve our goal.

Appendix A: Photographs

Uncontrolled state

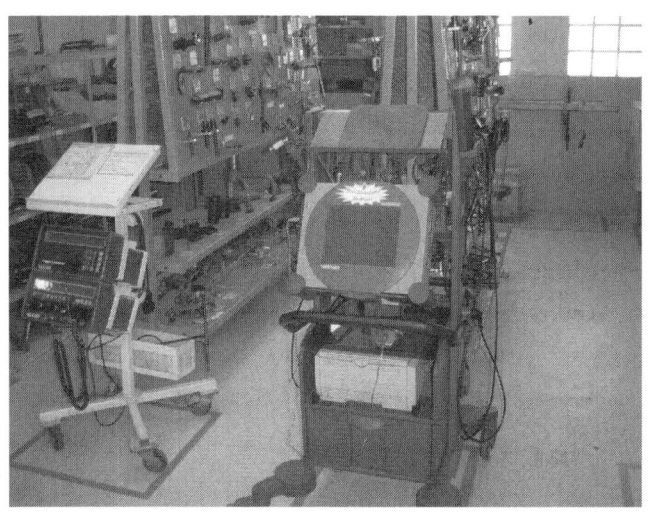

Benchmark condition: controlled 5S program

Appendix B: 5S Team Management

There are several stages of development that a team will most likely pass through on the way to becoming an efficient, productive group that can achieve the objectives set out at the start of any enterprise.

These stages can normally be grouped together into three main phases of development that the workgroup will pass through. These phases are:

- ➤ Immature Oriented Group (Chaotic Stage)
- ➤ Efficient Oriented Group (Formal Stage)
- ➤ Mature Integrated Group (Synergistic Stage)

Immature Oriented Group – Chaotic Stage

This is the initial stage where the group or organization has not yet been fully defined. The workflow tools that will ultimately be converted into schedules with resources for the work and its elapsed times and durations and the assignment of roles are not in place.

Leadership will be undefined, their talents and resources overlooked. There is little or no listening to others' views; some members may even withdraw from taking on active roles. Performance evaluation will not be conducted, mainly due to clear-cut performance goals not being set or considered. The needs of individuals will be overlooked; group members will tend to be self-centred, with only the loudest getting attention. Only when awareness of poor results and a clear task have emerged will the group move forwards to the next phase. The key skill at this stage for the leader is to diagnose the situation before acting.

Efficient Oriented Group – Formal Stage

This is the discipline-managed phase with a formal structure and a group leader. Communication will, in general, be improved, but group work is likely to be rigid. The workflow tools have been established with tasks defined and outcomes consistent and predicable to a large degree. Leaders exhibit directive leadership behaviour. While building knowledge and skills, the group starts to enjoy some success, but individuals have yet to identify their own needs. Some setbacks will give the group reasons to examine the causes and effects of those problems and become more probing and human in their approach to the enterprise. The group will a switch to a coaching style of management if the leader continues to direct and closely supervise the tasks but also explain his or her decisions. With this change, the group progresses into the next stage, where the needs of members and their tasks are blended to produce the optimum results. This stage of development can be considered even if the group has started to become a Mature Integrated Group.

Mature Integrated Group – Synergistic Stage

At this stage the skills of a group's members have been identified and drawn upon. The organization is flexible in its approach to work with self-discipline replacing the strong hand of the leader. The leader can take a supportive role and share the responsibility for making decisions through effective listening and learning. Group members express ideas openly, and the leader takes these ideas into account.

The workflow tools produce a working agenda for development of team members, and all tasks are identified and delegated. At this point, roles can be optimized and performance can continuously improve and the leader can use a delegating style, turning over responsibility for the day-to-day decision making to those team members who are highly competent and committed.

Summary

The maturity levels can be summarized with the help of the following two charts:

Maturity Level	Process Characteristic
Initial	Chaotic
Repeatable	Disciplined
Defined	Consistent
Managed	Predictable
Optimized	Continuously improving

Characteristics of an Immature versus a Mature organization

Characteristic	Immature	Mature
Over budget	X	
Late delivery	X	
Undefined processes	X	
Reactive	X	
Crisis management	X	
Poor quality	X	
Overworked, confused team members	X	
Unsatisfied customers	X	
Proactive, disciplined, consistent		X
Defined processes		X
Defined roles		X
Consistent monitoring		X
Predictive results		X
On time / within budget		X
Enabled team members		X
Satisfied customers (those downstream of the task)		X
Good communication		X
Processes institutionalized		X
Information systems viewed as a strategic function		X

Conclusion

➢ Progress through the stages is gradual and takes time.

➢ A group may fluctuate between the stages, and the leader will need to step backwards though the leadership styles with some team members

➢ Teams may never achieve the mature intergraded stage.

➢ Team-building training exercises can help teams develop.

➢ People develop at different rates, so a leader should not apply the same leadership style universally. Everyone has their own peak performance potential. A leader just needs to know how to tap into that potential.

➢ Effective communication is the key to bringing an enterprise and its team members to the optimized stage. Effective communication entails the promotion of the project to all its stakeholders through phase improvement – from being unaware, through the stages of aware and Interest and desire (attitude), to action (buying into the enterprise).

➢ A low maturity level can have a negative effect on an organization's ability to produce quality work.

Appendix C: Learning Organization

For programs such as 5S, TQM or any other, continuous improvement is required if their benefits are not to fall by the wayside. Continuous improvement requires a commitment to learning, while effective learning requires motivation, wanting to learn.

How can an organization improve without first learning something new? Solving problems and reengineering processes require us to see the world in a new way and then act accordingly. Without learning, organizations and people would simply repent old practices and repeat their mistakes.

Consider the following description of a learning organization: A learning organization is an organization skilled in creating, acquiring and transferring knowledge and at modifying its behaviour to reflect new knowledge and insights.

Foundations

➢ Problem solving – The organization uses a system which relies on the philosophy and methods of quality improvement, such as problem solving trees and fishbone charts and Deming's PDCA cycle for diagnosing problems. It uses data and simple statistical tools (histograms, P charts) as a background for making decision – fact-based management. Problem solving can be broken down into four steps 1. Understand the current situation 2. Identify the root cause of the problem 3. Make an action plan and 4. Carry the plan until the problem is solved.

Experimentation – The organization systematically searches for and the tests new knowledge. The use of scientific methods is essential to experimentation, but it differs from problem solving in that it focuses on seizing opportunities and expanding horizons rather than solving present problems. It normally takes two forms: 1) Ongoing programs aimed at producing small experiments to increase knowledge, and 2) pilot projects for testing out principles and processes that the organization may wish to use later on a larger scale. Both methods involve learning by doing.

All forms of experimentation move us from basic knowledge to a deeper understanding, from knowing how things are done to knowing why.

➢ Learning from past experience – A group uses of reflective observation to enable to learn from their experiences in which they review their successes and failures and systematically assess the lessons in a way that are open and accessible.

Reflection by itself, however, is not sufficient to promote learning after ten years of repeating the same experience. Unless a group acts on its reflections, it cannot develop. This process has been described as the 'Santayana review' after George Santayana, who said "Those who cannot remember the past are condemned to repeat it."

➢ Learning from others – Not all learning comes from refection and analysis. Insights from outside of a group's environment can allow it to gain new perspectives by providing fertile sources for ideas and creative thinking this technique can be applied to any process or function. To benchmark successfully, a group may need to employ a range of research techniques, including informal conversations with customers, employees, or suppliers; exploratory research techniques such as focus groups; or in-depth marketing research, quantitative research, surveys, questionnaires, reengineering analysis, process mapping, quality-control variance reports or financial ratio analyses.

➢ Transferring knowledge – For learning to be effective, knowledge must be distributed quickly and efficiently among the organization, as new ideas will gain maximum impact if they are shared widely in the organization. Education and training programs are excellent methods for transferring knowledge, but for them to be as effective as possible, the group must take action. Actively experiencing something is of far greater value than just having it described to you. You are more likely understand and retain knowledge if you have an easy transition to the new concept or practice by having the opportunity to apply it.

Measuring

The old adage, "if you can't measure it, you can't manage it" holds true for learning as with any other objective. For the most part, the solution to measuring learning has been to use learning curves. From these concepts emerged the half-life curve as a means to measure and compare improvements. The half-life curve measures the time required to achieve a 50 per cent improvement (with 50 per cent being a target of convenience). The logic is straightforward – the less time to improve halfway, the faster the learning. Shorter learning cycles translate to better performance.

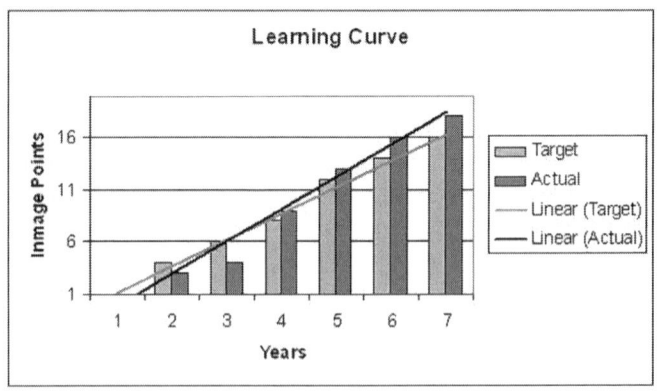

Customer satisfaction survey target and actual performance.
In the early stages, the learning rate was low.
By the fourth year, learning rapidly improved.

Half-life curves have an important weakness: they focus only on results. But some types of knowledge, such as TQM systems, take years to digest with few changes in performance for long periods. Because longer periods are required to obtain results, half-life curves are unlikely to capture the short-run learning. We should review learning in a more comprehensive manner to map progress.

Organizational learning can be mapped through three phases:

- ➢ Cognitive learning; when members of the organization are exposed to ideas, their knowledge is expanded and they begin to think differently.
- ➢ Internalization of these new concepts; members change their behaviour to account for these new insights.
- ➢ Tangible gains in quality, customer satisfaction, or another area.

Because cognitive and behavioural changes normally precede improvement in performance, we should review all three stages of the process to obtain a comprehensive understanding. Tools such as half-life curves or other measurements are needed to confirm that cognitive and behavioural changes have taken place and are producing results. Without these measurements, there is no rationale for investing in learning.

Rome was not built in a day; likewise, learning organizations are not built overnight. Most successful examples were the result of carefully cultivated attitudes, commitment and management processes.

The first step is to create an environment that encourages learning. There needs to be a time for reflection. Reflection by itself, though, is not enough to promote learning. Unless we act on those reflections, we will not develop. Learning from experience allows us to plan what to do next. Reflection is the central point that gives us the opportunity to draw conclusions about our practices. It is also called "abstract conceptualism".

From the conclusions we've drawn in the abstract conceptualism stage, we can then plan changes and move to the "active experimentation" phase, where the cycle starts again.

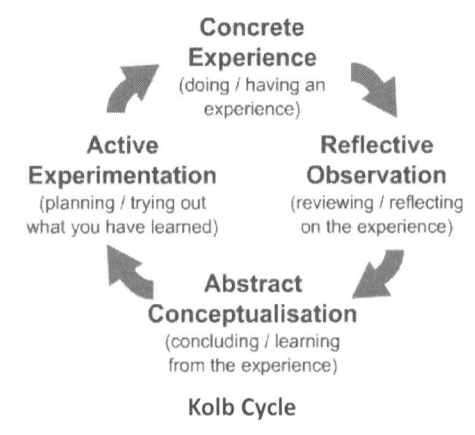

Concrete Experience
(doing / having an experience)

Active Experimentation
(planning / trying out what you have learned)

Reflective Observation
(reviewing / reflecting on the experience)

Abstract Conceptualisation
(concluding / learning from the experience)

Kolb Cycle

An important lever is the opening up of boundaries and the exchange of ideas. Boundaries prevent the flow of information and cause isolation and reinforce preconceptions. With a better understanding of the meaning, management and measurement of learning, we can create a learning organization.

References:

1. David A. Garvin, "Building a Learning Organization", *Harvard Business Review* (July–August 1993).
2. Clara Davies, *Kolb Learning Cycle Tutorial* (University of Leeds). http://www.ldu.leeds.ac.uk/ldu/sddu_multimedia/kolb/static_version.php

Appendix D:
Selected Reading and Resources

Material, Author (s), Publisher/Association, ISBN/Reprint

Thailand 5S Award 2009, Technology Promotion Association (Thailand – Japan), 5S Reference Guide, Resource Engineering, Inc., ISBN 1-88-230738-0

5S for Operators, Hiroyuki Hirano, The Productive Press Inc., ISBN 1-56-327123-0

ISO Made Easy, Kit Sadgrove, Kogan Page Ltd. London, ISBN 0-74-941275-5

ISO 9000 Guidance Notes, Consensus Books Sydney Australia, ISBN 0-73-374964-X

The Audit Skills Handbook, David Mallen and Christine Collins, Consensus Books Sydney Australia, ISBN 0-73-374793-0

The Definitive guide to marketing planning (edition published 2000), Angela Hatton, Pearson Education Ltd. Harlow, ISBN 0-27-364932-9

Production and Operation Management (5th Edition), Adam and Ebert, Prentice Hall Professional Englewood Cliffs, NJ, USA, ISBN 0-13-717943-X

5 Pillars of the Visual Workplace Published 1995, Hiroyuki Hirano, The Productive Press, ISBN 1-56-327123-0

5Ss: Five Keys to a Total Quality Environment, Takashi Osada, Asian Productivity Organization Productivity Press Inc. 1995, ISBN 9-28-331116-7

Balanced Scorecard 2002, Paul R. Niven, John Wiley & Sons, Inc. (20020 New York , ISBN 0-47-107872-2

The Balanced Scorecard , Nils-Göran & Anna Sjöstrand ,
Capstone Publishing (2002) Oxford UK, ISBN 1-84112-229-7

Balanced Scorecard in a Week , Mike Bourne & Pippa Bourne,
Hodders Headline London, ISBN 0-340-84945-2 978-0-
34094849-7

Understanding Total Quality Management in a Week, John
Macdonald, Hodder & Stoughton, ISBN 0-34-071191-4

Successful Project Management in a Week, Mark Brown,
Hodder & Stoughton, ISBN 0-34-070539-6

Building a Learning Organization, David A. Garvin, Harvard
Business Review, Reprint 93402

The Asia CEO in action, Korsak Chairasmisak, Direct Media
Group Bangkok, ISBN 9-74-913943-7

Shackleton's Way, Morrell & Capparell, Nicholas Brealey
Publishing Ltd., ISBN 10 1-85-788211-3

Delivering Quality Service, Zeithaml, Parasuraman, Berry, Free
Press, New York, ISBN 0-02-935701-2

Quality Assurance: Its Nature and Significance to the Motor
Vehicle Importer, E. B Moulding, Internal paper 1.06.1981,

Web Pages

Dartmouth
http://www.dartmouth.edu/~ogehome/CQI/PDCA.html

Health and Safety Executive
http://www.open.gov.uk/hse/hsehome.htm

Lean Manufacturing and the Environment
http://www.epa.gov/lean/thinking/kaizen.htm

Lean Manufacturing Strategy
http://www.strategosinc.com

MAS (Manufacturing Advisory Service)
http://www.mas.dti.gov.uk/content/resources/
categories/fact/FACT_7_wastes.html

The United Nations Industrial Development
Organisation (UNIDO) and the Japanese Standards
Association (JSA)
http://www.e4pq.org/tqm/index.php

http://www.thecqi.org/index.shtml

Acknowledgements

I was once told that the best way to learn is by speaking with others. This book represents conversations I have had with colleagues, friends and associates and the benefits I gained from the exchange of ideas and experiences. A book is never the work of one person – many people contribute to the ideas, encourage the author and shape the finished product. My thanks in this endeavour are due to:

Sumate Changrua, an automobile dealership's former General Manager, and Nupong Anukul, After Sales Manager, who successfully implemented the 5S program in their retail vehicle repair workshop.

Preecha Waragorngosol, Service Quality Manager, for translating certain material.

Dr. Paul Silva of Universidad Internacional Euroamericana.

Peter Burke for countless years of discussions.

20475139R00100

Printed in Great Britain
by Amazon